International Regulatory Co-operation: Case Studies, Vol. 3

TRANSNATIONAL PRIVATE REGULATION AND WATER MANAGEMENT

OECD
BETTER POLICIES FOR BETTER LIVES

This work is published on the responsibility of the Secretary-General of the OECD. The opinions expressed and arguments employed herein do not necessarily reflect the official views of the Organisation or of the governments of its member countries.

This document and any map included herein are without prejudice to the status of or sovereignty over any territory, to the delimitation of international frontiers and boundaries and to the name of any territory, city or area.

Please cite this publication as:
OECD (2013), *International Regulatory Co-operation: Case Studies,Vol. 3: Transnational Private Regulation and Water Management*, OECD Publishing.
http://dx.doi.org/10.1787/9789264200524-en

ISBN 978-92-64-20051-7 (print)
ISBN 978-92-64-20052-4 (PDF)

IWA Publishing
ISBN: 9781780405520 (Paperback)

The statistical data for Israel are supplied by and under the responsibility of the relevant Israeli authorities. The use of such data by the OECD is without prejudice to the status of the Golan Heights, East Jerusalem and Israeli settlements in the West Bank under the terms of international law.

Photo credits: Cover © Jörg Vollmer - Fotolia.com.

Corrigenda to OECD publications may be found on line at: *www.oecd.org/publishing/corrigenda*.

Foreword

This report is part of a mini collection of books on the topic of international regulatory co-operation (IRC). It comprises two case studies, upon which the synthesis report (*International Regulatory Co-operation: Rules for an Interdependent World*) builds:

- Transnational private regulation;
- Transboundary water management.

These case studies have sought to capture the main characteristics of selected IRC experiences and follow a common structure to ensure comparability of approach.

This work on IRC has been conducted under the supervision of the OECD Regulatory Policy Committee whose mandate is to assist both members and non-members in building and strengthening capacity for regulatory quality and regulatory reform. The Regulatory Policy Committee is supported by staff within the Regulatory Policy Division of the Public Governance and Territorial Development Directorate.

The OECD Public Governance and Territorial Development Directorate's unique emphasis on institutional design and policy implementation supports mutual learning and diffusion of best practice in different societal and market conditions. The goal is to help countries build better government systems and implement policies at both national and regional level that lead to sustainable economic and social development. The directorate's mission is to help governments at all levels design and implement strategic, evidence-based and innovative policies to strengthen public governance, respond effectively to diverse and disruptive economic, social and environmental challenges and deliver on government's commitments to citizens.

This publication was co-ordinated by Céline Kauffmann, Senior Economist, under the supervision of Nick Malyshev, Head of the OECD Division on Regulatory Policy. The case study on transnational private regulation was written by Fabrizio Cafaggi, Andrea Renda, and Rebecca Schmidt in the framework of the Hague Institute for the Internationalisation of Law Project. The case study on transboundary water management was written by Julia Black, London School of Economics and Political Science, and Celine Kauffmann, OECD, with inputs from Aziza Akhmouch, OECD, and research assistance by Flavia Donadelli, London School of Economics and Political Science. The report was prepared for publication by Jennifer Stein.

Table of contents

Figure

Abbreviations and acronyms

ANSI	American National Standards Institute
ARSO	African Organization for Standardization
CEN	European Committee for Standardization
CFA	Co-operative Framework Agreement
COPANT	Pan American Standards Commission
DIN	Deutsches Institut für Normung
FIFA	Federation International de football Association
FLO	Fairtrade Labelling Organizations International
FSC	Forestry Stewardship Council
GFSI	Global Food Safety Initiative
GRI	Global Reporting Initiative
ICANN	Internet Corporation for Assigned Names and Numbers
ICPR	International Commission for the Protection of the Rhine
IEC	International Electronic Commission
IFOAM	International Federation of Organic Agriculture Movements
ILO	International Labour Organization
INGOs	International non-governmental organisations
IOAS	International Organic Accreditation Service
IOC	International Olympic Committee
IRC	International Regulatory Co-operation

ISO	International Organization for Standardization
ITU	International Telecommunications Union
LEI	Lembaga Ekolabel Indonesia
MAC	Marine Aquarium Council
MoU	Memorandum of Understanding
MSC	Marine Stewardship Council
PEFC	Programme for the Endorsement of Forest Certification
RAP	Rhine Action Programme
SAAS	Social Accountability Accreditation Services
SAI	Social Accountability International
TPR	Transnational Private Regulation
UEBT	Union for Ethical BioTrade
W3C	World Wide Web
WMO	World Meteorological Organisation

Chapter 1

Transnational private regulation

by

Fabrizio Cafaggi, Andrea Renda and Rebecca Schmidt*

As markets and regulatory tasks become increasingly global, forms of private international regulatory co-operation are emerging along with – or sometimes as a replacement for – inter-governmental co-operation. In a number of settings, traditional forms of public intervention are facing enormous, sometimes insurmountable difficulties in coping with certain policy problems. The weaknesses of public regulation emerge more specifically at the transnational level where difficulties to co-ordinate, inconsistencies between standard setting and enforcement, divergences between administrative and judicial enforcement and within the latter among domestic courts make inter-state regulatory co-operation an insufficient response. This case study analyses how the development of transnational private regulation responds to the needs of globalisation, while raising a number of challenges.

* Fabrizio Cafaggi is professor of comparative law and HiiL Morris Tabaksblat Visiting Chair on Private Actors and Globalisation at EUI/RSCAS. Andrea Renda is HiiL Morris Tabaksblat Visiting Chair on Private Actors and Globalisation at EUI/RSCAS. Rebecca Schmidt is PhD Candidate at the Department of Law at the European University Institute.

The context of transnational private regulation

As markets and regulatory tasks become increasingly global, forms of transnational regulatory co-operation are emerging along with – or sometimes as a replacement for – inter-governmental co-operation. In a number of settings, traditional forms of public interventions are facing enormous, sometimes insurmountable, difficulties in coping with certain policy problems. The weaknesses of public regulation emerge more specifically at the transnational level where difficulties to co-ordinate, inconsistencies between standard setting and enforcement, divergences between administrative and judicial enforcement and, within the latter, among domestic courts, make inter-state regulatory co-operation an insufficient response.

As a result, transnational private regulators have emerged. They often pursue public interest functions also due to the multi-stakeholder composition of the regulatory bodies. This case study examines forms of transnational co-operation and competition between private regulators and modes of interaction between public and private actors at international level. The first section provides a short overview of transnational private regulation (TPR), addressing the five conventional questions in a comparative fashion: why, who, what, how and where. Section 2 outlines the current evolution of co-operation between TPR schemes as well as between TPR and other schemes. Section 3 then provides an assessment of the current state of TPR within the broader context of international regulatory co-operation, with a focus on the need for more complete and rigorous indicators for assessing the effectiveness of TPR schemes. Section 4 concludes and highlights future steps.

Main characteristics of transnational private regulation

Why is TPR taking place?

There are numerous reasons why TPR regimes are established, some are of more general nature, others field-specific (e.g. private regulation regarding food safety, forestry management or derivates). Four major factors can be seen as having caused and helped to accelerate the emergence of TPR schemes.

First, many goods and services today transcend national boundaries and as such are not easily regulated by national legislation. Global trade and governance of inter-state externalities can hardly be regulated through domestic regulations. This is particularly the case with global public goods, (e.g., deforestation, emission reduction, protection of biodiversity, food safety, financial stability), for which international regulatory co-operation is substantially needed to avoid "race to the bottom" between domestic regulations. Traditional international law instruments often fail to provide responses due to a lack of political consensus. This often triggers the emergence of transnational private regimes. Early experiences such as the Forestry Stewardship Council (FSC) and the Marine Stewardship Council (MSC) have filled the governance gap created by failed international regulatory co-operation. Only at the end of the last century and, more intensely in recent years, the global policy community has become greatly aware of the need to strengthen regulatory co-operation, in particular through the work of the OECD in its initiative on regulatory co-operation, by the UN in the field of corporate social responsibility, by FAO together with a number of established meta-private regulators such as the ISEAL alliance.[1]

Second, when good governance standards are met, TPR can contribute to promoting the growth of regulatory capabilities and compliance with the rule of law. There are global value chains that extend to countries in which the rule of law is not entirely complied with, where contractual governance may partly replace public domestic institutions and contracts are used as regulatory instruments. The use of supply chains as instruments of transnational regulatory co-operation is an attempt on the one hand to improve regulatory effectiveness, where domestic regimes do not have strong regulatory states. On the other hand, it reduces transaction costs related to regulatory diversity in the public domain. Contractual networks among suppliers that aim at ensuring the respect for minimum standards under the supervision of large MNCs and NGOs are one illustration of such changes. This form of governance adopts mainly hierarchical instruments with a chain leader where one (large) retailer or several retailers and/or producers set, often social, environmental or product safety standards to be implemented throughout the supply chain.[2] These forms may lack legitimacy and be under-inclusive.

Third, there are markets that exhibit very fast-changing dynamics – so fast-changing that it is difficult for public policy makers to try to regulate them. In particular, this is the case of high-tech and knowledge-intensive markets, in which the fast pace of change and the highly technical nature of the information needed to regulate effectively leads policymakers to rely on private parties, at least for the definition of implementing measures and technical specifications. Examples include, but are not limited to, international roaming, net neutrality regulation, cloud computing, privacy on the Internet, and Internet governance. Bodies such as the Internet Corporation for Assigned Names and Numbers (ICANN),[3] the Internet Engineering Task Force or the World Wide Web consortium (W3C) are good examples of private regulation in this field, whereas the UN-sponsored Internet Governance Forum is an example of a multi-stakeholder platform where public and private players interact.

Last, there are policy problems that inevitably require heavy reliance on the expertise of private actors, since the latter are the most informed parties, or the players in the best position to deal with a given failure, or simply the only parties holding control over essential resources. These fields include widely diverse policy domains, such as, new technologies as addressed in the last paragraph, as well as generally technical standards, sustainability reporting, critical infrastructure protection and many others.

These trends suggest that transnational private actors have resources, expertise and capacities that can complement the role of traditional international organisations and the new forms of regulatory co-operation such as transnational networks. TPR is not an alternative but rather complements public international regulation.

Who are the transnational private regulators?

TPR is driven by multiple actors: Firms, NGOs, independent experts, or epistemic communities (Cafaggi, 2011a). The importance of NGOs constitutes a distinctive feature of TPR reflecting the transformations within the private sphere. They have shifted from being rule-takers and final beneficiaries of regulatory processes to rule-makers. Private actors often express divergent and, at times, conflicting interests marking deep differences with the more traditional domestic forms of self-regulation where a higher degree of homogeneity exists.

Private actors adopt different organisational typologies: Single stakeholder (organisations) representing industries or NGOS; multi-stakeholder organisations comprising various categories of actors and/or memberships. Actors' involvement depends on the objectives pursued by the regulatory entity and the concentration of regulatory power. Firms have a substantial interest in getting involved into regulatory frameworks that facilitate their business practices and expand their reputation and market shares, such as food safety, financial markets regulation, or accounting standards. NGOs are involved in regulation that concerns their area of expertise or interest representation, such as labour rights, environmental protection, protection of certain groups of society (indigenous people, minorities etc.). They have incentives to maximise their influence on regulatory processes on behalf of their constituencies.

Apart from regulating core business activities or the specific areas of advocacy, the evidence shows broader engagement of private actors in expanded areas of regulation. For example, due to changing market and regulatory conditions, firms are becoming increasingly involved in private environmental and social regulation. NGOs, on the other hand, might see an interest in co-operating with industry in areas where they would not have a strong impact using adversarial models, such as the Marine Stewardship Council (MSC), born out of a co-operative project between Unilever and the WWF. Industries and NGOs represent conflicting interests and divergent policy objectives. Increasingly however they engage into transnational co-operation rather than private regulatory competition. As a result multi-stakeholder organisations are replacing many single-stakeholder regulators.

Even within the private sphere expertise-based regulation is becoming more common with significant delegation to technical standard setting bodies whose scope has grown tremendously over time. A significant role is played by technical standard-setting bodies, often private in terms of legal form, but public or semi-public in terms of functions performed and regularly monopolistic in their area of activity (Büthe and Mattli, 2012). There is a large number of technical standard setters. They are often specialised in a specific area. They can be either public or private, depending on their founding documents. Among the bigger, international and long-established ones, one finds the International Organization for Standardization (ISO), which is privately incorporated in Switzerland, the International Telecommunications Union (ITU), a public organisation founded by treaty, as well as the International Electronic Commission (IEC), the World Wide Web Consortium (W3C), the Codex Alimentarius Commission, an intergovernmental institution, just to name a few. ISO, ITU and IEC form the World Standards Cooperation Alliance, a co-operative approach that is supposed to strengthen the voluntary standard setting regimes.[4]

International organisations often take the form of a federation with a multilevel structure based on national bodies. In fact ISO members are the respective national standard setters, such as the American National Standards Institute (ANSI) in the USA, Deutsches Institut für Normung (DIN) in Germany, or the British Standards Institution (BSI). The picture is completed by numerous regional technical standard setters that provide for harmonised technical standards within a specific region. Examples are the European Committee for Standardization (CEN), the Pan American

Standards Commission (COPANT), the African Organization for Standardization (ARSO), and many others. Technical standards are progressively expanding towards areas that were not initially conceived as technical.[5] This process reflects the use of technical standardisation and standardisation bodies as tools to find compromises among different private interests, both within industries of developed and developing countries and between industry and NGOs.

What is subject to regulation?

TPR initially developed as sector-specific regulation. The environment, private security services, financial markets, technical standards, e-commerce, food safety, data protection, e-commerce, etc. are just some areas in which private regulators are active. As TPR schemes vary enormously in terms of composition, mission, geographic reach and governance structures, it comes as no surprise that their specific regulatory functions and objectives can change. There are however paths of policy integration: what used to be a contested field between environmental and social standards has become an integrated one. Many transnational regulatory instruments take an integrated approach regulating contextually environmental risks and social standards so as to avoid unnecessary trade-offs. Another typical example is conservation and land trusts, or forms of regulation by contract that aim mostly at fostering compliance where litigation would prove costly, unpredictable or simply impossible.[6]

A good way of looking at the scope of TPR schemes is to highlight the stage of the policy cycle in which they are involved. Some forms of private regulation mostly focus on the setting technical standard setting, as is the case for ISO and other private technical standard setters. Other organisations help implement existing international public soft and hard regulation.[7] An increasing number of private regulators focus on monitoring and compliance. This is particularly the case with certification bodies, whose main function is to ensure and certify that various types of public and private standards are complied with (Schmidt and Verbruggen, 2013). Often several private regulators occupy the regulatory space leading to fragmentation. More rarely, there are some types of organisations that act with undisputed authority in a given policy field.[8]

More recently, meta-private regulators have emerged to provide general rules that are functionally applicable to many sectors. This is the case for reporting where there are general schemes, specified for each sector; and, more recently, regarding the publication of general codes concerning principles for standard setting, monitoring and compliance, aimed at a wide range of organisations.[9] This trend suggests a pattern towards the creation of common rules and principle cutting across different sectors favouring co-ordination without leading to organisational integration.

How do these organisations regulate?

The regulatory instruments deployed in TPR are primarily drawn from private law, in particular ownership and agreements, property rights and contracts. Private standards are mainly voluntary standards that regulated entities freely decide to adopt. However, legally voluntary often becomes *de facto* mandatory when market or social factors force regulated entities to adopt them. In addition, a significant number of private standards are incorporated into legislative or administrative acts and become binding. Consequently, the assumption of voluntariness should be qualified by considering many legal and non-legal factors that reduce or eliminate the space of choice for regulated entities.

When looking at co-operative forms, the two main venues of TPR are organisations and contracts. Organisations can have various forms (association, foundation, for profit, not for profit, etc.). First and foremost, they regulate the behaviour of their members, but they can also have impacts on third parties. As in the field of public international regulation, global private regimes often take the form of transnational networks but tend to be more formalised than their public counterparts.

Agreements and contracts are the dominant tools. To the traditional codes of conduct and guidelines, commercial contracts have been added as a means of ensuring compliance giving enforceable rights to supply chain participants, more rarely to consumers, consumer, environmental or human rights organisations. Through such transnational contracts, regulation created by organisations can be imposed on third party actors along the chain across state boundaries. The fastest growing phenomenon is the use of supply chain as regulatory vehicles (Cafaggi, 2013). Frequently, buyers at the retail stage define the regulatory regime themselves or incorporate rules into the contracts with spillover effects on the whole supply chain and impose compliance with those rules along the chain. An example is the M&S CO_2 reduction target that involves reducing energy consumption by 10% in its top 100 clothing factories (OECD, 2010). The new instruments tend to be binding relying on both legal and non-legal enforcement mechanisms.

Where does regulation occur?

TPR is multilevel and often combines forms of territorial and functional integration. From the territorial perspective, there are instances where local regulators create a global entity and others where a global regulator is created first and then local/national chapters are formed. TPR does not necessarily follow the administrative partitioning of public entities. In particular, there is no necessary coincidence between the territorial scope of regulation and the nature of the regulator. Global regulators, such as global supply chains, become involved in regional regulation when market and institutional reasons so require. Along the same supply chain, private regulation can affect different segments corresponding to regional regimes.

Evolution of regulatory chains: The birth of meta-regulators

Fragmentation characterises many private regimes. Forms of transnational co-operation have to be supported by common rules, especially when co-operation is multiparty, involving a large number of schemes. A trend can be observed towards the creation of meta-organisations and meta-standardisation as a way to bridge the gap between private regulatory schemes and public policy makers. Facing strong fragmentation, IGOs and domestic regulators call for simplification and ask for shared rules that can be more easily implemented thereby increasing effectiveness.

An important example in this respect is the work of ISEAL Alliance, an international non-profit organisation that codifies best practices for the design and implementation of social and environmental standards initiatives.[10] This is a landmark example of a "private meta-regulator". Besides a Standard-Setting code, recently the ISEAL Alliance launched an important tool for the purpose of evaluating external impact of private regulatory bodies. The Code of Good Practice for Assessing the Impacts of Social and Environmental Standard ("Impacts Code") sets out the process by which standards systems can prove their contributions to social and environmental impacts as well as learning about, and improving, the effectiveness of their system. The Code appears strikingly similar to an Impact Assessment guidance developed by public policy makers.[11] The last code is the assurance code dealing primarily with enforcement questions. These codes are intended for private regulators which are members of ISEAL but have been referred to, and informally endorsed, by many public international organisations.

The development of an Impacts Code filled the lack of a suitable definition of sustainability. While the ISEAL Impacts Code primarily looks at the stakeholder in a given private governance scheme (e.g. the MSC, FSC, RSS), the conceptual framework definition provided by ISEAL aims at representing a "universally applicable definition of sustainability" (Guttenstein, et al., 2010). This development led to the involvement of previous voluntary certification systems (FLO, MSC, UTZ Certified); tools generated by UN bodies (UN/ECOSOC, FAO, ILO, UNEP, etc.); corporations (e.g. Wal-Mart Sustainability Index); NGOs (e.g. Transparency International, the Bellagio STAMP); and academia (e.g. the Stiglitz-Sen-Fitoussi Report by the Commission on the Measurement of Economic Performance and Social Progress). Within this context, a definition of sustainability has emerged, which can be currently considered as being among the most sophisticated and widely acknowledged definitions to be used for public policy purposes.

The ultimate consequence of this development has been the full endorsement of ISEAL's efforts by the FAO together with other private instruments, which has now taken the leadership in the definition of a global Sustainability Assessment of Food and Agriculture systems (SAFA), notably through the drafting of guidelines.

An overview of transnational regulatory co-operation

The move from sector-specific to general private regulation and the development of a consistent institutional design are also aimed at facilitating co-operation among potentially conflicting private organisations which strive to gain power in a contested regulatory space. Two concurring phenomena can be observed: *i)* organisational integration via mergers or the creation of new regulatory entities; *ii)* contractual co-operation among organisations representing different interests or operating in conflicting policy fields. The former reflects the goal of a long-term, general co-operative aim; the latter tends to be project-specific and may be short- or medium-term. In some instances contractual forms constitute the first stage of a process to integrate organisations often after pressure coming from IGOs and external competition.

In this context, the parallel co-operation between public and private is very interesting, giving rise to hybrids which deploy a combination of instruments drawn from domestic private and administrative law. Public entities play a significant role in steering transnational private regulatory co-operation calling for less fragmentation and adopting forms of mutual recognition or even consolidation schemes. It is probably too early to say whether, in turn, a new era of co-operation between public and private players will come from the growing consolidation and streamlining of TPR schemes and the emergence of private meta-regulators. But the examples outlined below can clearly set the stage for enhanced integration and knowledge sharing between players participating in various schemes at inter-governmental and private level. The following section provides an overview of the most relevant aspects of transnational regulatory co-operation. This includes first a typology of the actors involved, followed by a detailed description of how such co-operation is framed. Finally, the section looks into the reasons for which the various actors become involved in regulatory co-operation.

A typology of the actors involved

The actors of TPR – typically firms, NGOs, independent experts or epistemic communities – choose among different organisational models to engage into regulatory co-operation which affects processes and outcomes. The selection of governance forms influence the regulatory outcomes. Inclusive governance tends to produce more principle-based regulation leaving detailed specifications for a later stage. Similarly to single regulators, the forms of regulatory co-operation partly reflect those highlighted previously in relation to single regimes. The main difference is the greater use of agreements and memoranda of understanding and networks. In relation to regulatory co-operation, a distinction can be made between single stakeholder (organisations), representing industries or NGOs; multi-stakeholder organisations, including different actors and/or memberships categories; and technical standard-setting bodies, private in terms of legal form, but public or semi-public in terms of functions performed. The single stakeholder co-operative forms concern co-operation among enterprises of different sizes and locations or different NGOs. On the other hand, the multi-stakeholder reflects a form of transnational co-operation among regulators led by different constituencies.

To provide a classification, distinction between private and public actors is made on co-operation. The distinction is relevant because it reflects the different sources of regulatory power, necessary for engaging into regulatory co-operation. In intergovernmental organisations, the source of regulatory power is the contracting states; in private regimes the source is private autonomy, which may be consensual or hierarchical depending on the distribution of market power. However, it is not always easy to distinguish between public and private. "Private organisations" can include public members or even fulfil a public mandate.[12] Public entities might have a high level of private actor participation and might also be influenced significantly by the latter when defining their goals and even in the instruments' choice.[13] To avoid confusion, a formalistic distinction based on the founding document and the sources and instruments of regulatory co-operation (treaty, charters or by-laws, contracts or agreements) is adopted. Accordingly, an organisation such as ISO, where national standard-setting bodies co-operate to create global standards, should be considered a private entity since it is an association under Swiss law. On the other hand, the United Nations Global Compact will be grouped as public, since it is a policy initiative launched by the UN despite the fact that the majority of its participants stems from business.[14] It is evident that this is an oversimplification since a functional approach would reshape the boundaries underlining the mutual influence played by public and private in the light of the institutional complementarity approach. However, strong legal implications cannot be overlooked arising from the public or private nature of the organisation and the instrument of co-operation.

Nonetheless, this shows that TPR is not limited to private actors. Many forms of TPR include public bodies that agree to operate with typical private law instruments like associations, agreements and contracts. When TPR includes public bodies, additional actors could be involved such as traditional IOs founded by a treaty (typically the WTO, the WHO, the UN, and the OECD), supra-national entities (in particular the European Commission), networks (such as the UN Compact or the World Commission on Dams) and single states, or sub-level state entities (see for instance the engagement of the US Environmental Protection Agency).

Figure 1.1. The governance triangle

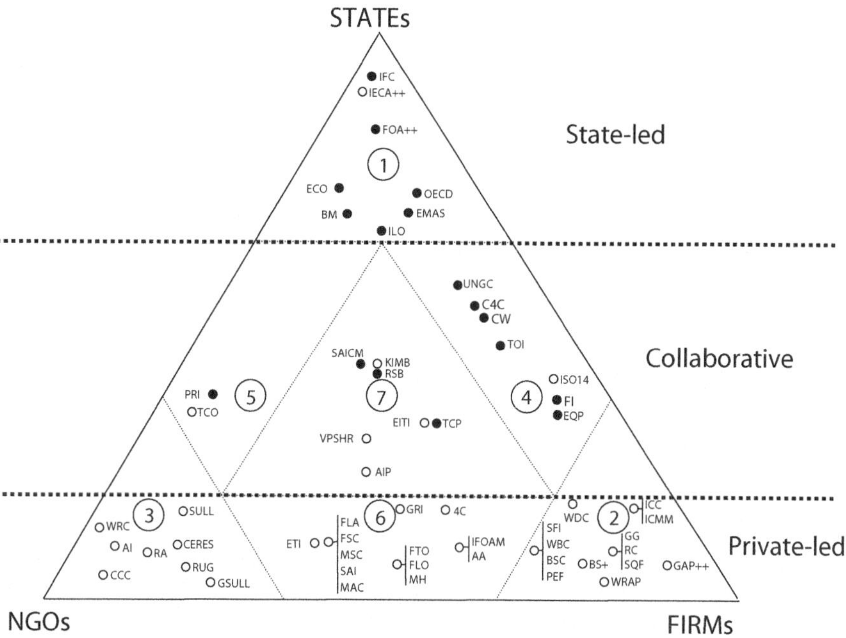

Source: Abbott, K. and D. Snidal (2010), "International Regulation without International Government: Improving IO Performance through Orchestration", Vol. 5, *Review of International Organizations*, p. 315.

Abbott and Snidal have tried to reach an operational taxonomy by distinguishing existing schemes based on the nature of their participants. Figure 1.1 shows their "governance triangle", in which a large number of governance organisations are located along a triangular space based on the relative prevalence of State actors, NGOs or private firms. This allows them to identify as many as seven different "zones", corresponding to different mixes of state, NGO and business participation. Vertex zones in the triangle (1-3) correspond to organisations with single actors or sets of actors, with limited cross-group participation and thus limited multi-stakeholder governance. For example, Zone 1 can include the OECD Guidelines for Multinational Enterprises or domestic/inter-governmental labour law agreements; Zone 2 includes firm and industry self-regulatory schemes; and zone 3 includes codes promulgated and administered by NGOs and NGO coalitions. As further described by Abbot and Snidal, Zones 4-6 include schemes in which actors from two groups share governance responsibility (e.g. zone 4 contains the UN Global Compact, in which civil society only plays a small role. Zone 7, the central triangle, includes institutions in which actors of all three types play a vital role – for instance ILO's Declaration on Multinational Enterprises and Social Policy (Abbot and Snidal, 2009). Their analysis confirms a trend towards the multi-stakeholder model.

Another important differentiation is the regulatory nature of the actors involved. A distinction is made here between two groups operating at transnational level: regulatory organisations and entities whose primary objective is not regulation.

- *Entities whose primary objective is not regulation* (business actors[15] and NGOs,[16] etc.), but who are incidentally pursuing a regulatory agenda in the area they are co-operating in. This often occurs when multinational companies conclude (multiparty) contracts that have primary commercial objectives but indirectly have a regulatory impact. Similarly to regulatory organisations, those entities might create or participate in regulatory organisation or use contractual tools for their co-operative projects.

- *Regulatory organisations,* i.e. formally established organisations (e.g. associations, foundations or corporations),[17] co-operate either via organisational form or simply through regulatory agreements. In contrast to other entities such as business actors or NGOs, whose main purpose is profit making or political advocacy, regulatory organisations have predominantly regulatory activities. Again borders are not easy to draw. Regulatory entities may also follow business objectives,[18] as well as other organisations, particularly in the environmental or social realm, are often considered NGOs.[19] Regulatory co-operation among private entities can therefore be driven by profit motives or public interest reasons giving rise to very different regimes especially concerning the effects of co-operation on third parties.

How do TPR schemes frame and implement co-operation?

As mentioned above, TPR schemes are regulated in various forms, depending on their ultimate mission and objectives. The most common forms of regulatory output produced by TPR schemes are codes of conduct and guidelines, and also commercial contract schemes, industry standards and social or environmental standards. New instruments tend to be binding and move away from the voluntary nature which has characterised early years. A similar toolkit is used to engage in regulatory co-operation. The following section analyses the forms of private regulatory co-operation and addresses: first the various levels at which actors co-operate, followed by a description of the degree of formality; then the various instruments of co-operation; and finally the functions that are being co-ordinated.

Levels of co-operation

When looking at regulatory co-operation in the private domain, two main levels can be distinguished. On the first tier, different actors come together to carry out a common regulatory project (short-term, long-term or permanent). This is the case with most regulatory organisations, which are usually built up with different actor groups.[20] It can, however, also be carried out through agreements, such as by two corporations that sign a contract for specific regulatory objectives. The second tier constitutes a meta-level where the objective is to regulate the regulatory process. Here again, regulatory organisations, which are already established as such, co-operate either to pursue a specific regulatory agenda (e.g. creating the ISO 26000 Guidelines), or to create rules for their general activities (as it is the case with ISEAL Alliance, which sets rules for regulatory activities).

An important form of regulatory co-operation is represented by mutual recognition. Increasingly, schemes formerly competing move from competition to co-operation designing common equivalence rules that permit regulated entities to select one regulator without incurring risks of duplicating costs of compliance. A good illustration is the global food safety initiative (GFSI) in the field of food safety. Functional equivalence assumes a significant level of common rules permitting differentiation of means to achieve the same objectives. Mutual recognition could not be reconciled with divergent objectives. A lower level is pursued with compatibility. Here the goal is to make regimes that have divergent objectives compatible with one another.

Formality

The adoption of TPR is voluntary but its compliance is often binding on the signatories (Caffagi, 2011). It is often mistakenly assumed that voluntariness and informality coincide. There is only a subset of regulatory co-operation instruments, which are voluntary and not legally binding; they are referred to as informal. For the rest it is assumed that once adopted they are legally binding and enforceable.

Moreover, informality should not be linked with less effectiveness and legitimacy. Empirical research shows that informal transnational regulatory co-operation, even when it deploys non-enforceable agreements or MOUs, may be highly effective and *de facto* is perceived as binding. Legal enforceability by the courts is only one of the factors that make a co-operative agreement effective. Often social, economic and political factors have a greater influence on compliance than legal enforceability providing signatories with the incentives to comply.[21]

As a general rule, when an agreement is defining a concrete regulatory project there is tendency to use binding provisions. As mentioned, the MoU between ILO and ISO very strongly stipulates the need to ensure consistency between ILO standards and ISO 26000 (in particular, see ILO and ISO, 2005, Article 1). Agreements concerned with technical details of a co-operative venture also show a higher tendency towards hard language, since their success depends very much on the parties complying with the specific provisions (see IEC, ISO, ITU and UN-ECE 2000; SOCHI 2014 and UNEP 2009). On the other hand, when an agreement is intended to establish a new regulatory relationship, before crystallising the concrete steps of the project, the wording is softer and parties intend to make a political rather than a legally binding commitment. Examples for such agreements are the MoUs between GC/EFQM and between UNEP/FELABAN. However, in most cases one finds both provisions. Only very rarely is a text entirely drafted using "soft provisions". A particular exception in this regard is the MoU between UNEP and the US Environmental Protection Agency, which only uses wordings such as "*intend*" or "*may*".

Moreover, the involvement of national or European regulators in TPR schemes may often be informal. Often public involvement in private regimes is used to facilitate certain agreements, to provide advice, to achieve voluntary adoption or to provide a platform for co-operation.[22] However, this public involvement may also take more concrete and binding forms. In the accounting field, the Commission is participating in IASB deliberations through the European Financial Reporting Advisory Group.[23] Standards are then introduced in the EU legal order via a formal Commission decision. In the area of civil aviation, national aviation authorities and Eurocontrol are members of the European Strategic Safety Initiative, a voluntary, non-legally binding and privately funded programme initiated by ICAO, a specialised agency of the UN.

Instruments used for co-operation

The two main instruments used for transnational regulatory co-operation are organisations and agreements. Co-operation in organisational form can either be on a membership basis, when either the partners become members of the respective other organisation(s) or when a new organisation is founded. Granting liaison status within the other organisation would be a less formal venue. Finally, the network form is used to engage in softer forms of co-operation that do not require the creation of a new entity with legal personality. Co-operation via agreement is a very broad category, which encompasses various typologies, from formal and legally binding regulatory co-operation agreements, to commercial contracts, and finally to soft law arrangements.

Agreements (often in the form of MoUs) are frequently used both in the public and private domains. Government agencies use them to co-operate with each other or with third parties, such as international organisations and private entities.[24] Private actors use it either to co-ordinate their regulatory schemes,[25] or to start new regulatory projects.[26] In between, one finds a wide variety of interaction between public and private or other hybrid forms of regulators based on agreements.[27] As already mentioned above, agreements are difficult to group into particular types since they can govern many different forms of regulatory co-operation. Nonetheless some very broad parameters for grouping regulatory agreements have been identified: *i)* the time frame used, *ii)* the inclusion of procedural and substantive standards *iii)* single or multiple projects.

Even though many agreements have a particular focus, they usually encompass numerous regulatory projects,[28] and only a few are concerned with one single project.[29] Regarding duration, there is equal distribution between agreements concluded for a limited time frame and those set for an indefinite one. It seems that a fixed time frame is chosen either when a particular project is at stake or when the agreement is a first step in establishing co-operation.[30] The documents usually include both substantive as well as procedural provisions. Even rather vague MoUs aimed at potential co-operation contain both substantive and procedural provisions. The typical MoU has a substantial part, which is usually the declaration of "purpose", where the co-operative project is outlined and often some more specific provisions are also included. Following this, a number of procedural aspects are discussed.[31] Generally, all kinds of combinations between procedural and substantive norms are possible. Some agreements include a higher degree of procedural norms, such as the agreements between ISO and the respective IOs in the context of ISO 26000. Other agreements are somewhat more substantive. Finally, a significant number of provisions cannot be clearly classified into either category.[32]

Finally, the various forms of regulatory co-operation are by no means mutually exclusive. Obviously, the founding document of a private regulatory organisation is an agreement entered into between the members. Accession to such an organisation is governed by agreement and even less formal forms such as liaison participation is in written form.[33] On the other hand, agreements can include participatory aspects or even set forth certain forms of organisational co-operation.[34] Often a combination of organisational and contractual instruments is used to govern more advanced forms of co-operation.

Functional dimension of co-operation

Preliminary stage

Within regulatory agreements dealing with co-operation, a number of pre-regulation tasks interlinked with the regulatory process are often defined. In particular, these are information exchanges, as well as the definition of common goals, agendas and strategies. Exchanges of information are included almost in all private co-operative projects. They are to be found not only in agreements but also among organisational rules.[35] Common goals, regulatory agendas and strategies are usually defined at early stages of a co-operative project.[36]

Co-operation within the regulatory process

The regulatory process is usually divided into three stages (standard setting, monitoring and enforcement).[37] Co-operation can and does occur at all three levels. Nonetheless, it is often difficult to clearly distinguish between the different stages.

In organisational form, co-operation often covers all the three regulatory stages (at least to varying degrees). Some organisations have very encompassing systems that include standard setting, monitoring, as well as enforcement or at least compliance mechanisms. Examples of this are organisations that work with certification schemes such as the FSC, the MSC or GlobalG.A.P. etc., which usually set standards and then, in one form or another, also co-ordinate the certification process.[38] Other organisations are more focused on the standard-setting processes themselves[39] and either leave the remaining stages to other actors (entirely independent certifiers, market participants, or the public etc.),[40] or sometimes do not even anticipate them.[41]

Co-operative projects based predominantly on agreements are often concerned with the harmonisation or customisation of standards originating from various institutions but regulating overlapping areas.[42] Often the agreement provides for comprehensive integration of the three stages: standards have to be harmonised or integrated (see MoU between GRI/GC), and furthermore, monitoring and enforcement activities (or at least compliance procedures) are adapted at the same time. Concretely, many agreements contain provisions regarding the promotion, strengthening, or implementation of regulatory standards to increase overall compliance by way of co-ordination (see the provisions of the MoU between LEI/FSC).

Activities surrounding the regulatory process: Recognition of standards – benchmarking

Benchmarking regimes and other forms of recognising or harmonising existing standards are a form of co-operation aimed at harmonising standards and facilitating compliance with them. Benchmarking systems are often found in areas where a great number of regulators are active, such as food safety or forestry protection. Benchmarking systems can come with membership and often include some degree of peer review.[43]

National and European public regulators sometimes also provide for the recognition of private standards through their own regulatory frameworks. This may be seen in several environmental regulatory regimes. In the Biofuels Directive, the European Commission recognises privately set schemes that verify compliance with the public sustainability criteria.[44] The so-called Timber Regulation provides for certification schemes that can be applied for due diligence when placing timber products on the market.[45] Another example is accounting, where Regulation 1606/2002 requires all EU listed companies to apply the IFRS for (see "Regulation on the Application of International Accounting Standards", No. 1606/2002/EC, 19 July 2002). In return the EU participates in the standard-setting process through the European Financial Reporting Advisory Group.[46]

Why are co-operative TPR schemes created?

The origins of co-operation between private governance schemes can be related to various causes and motivations. As outlined above, TPR is often built on co-operation between different actors (first tier). In addition, many already established private regulators co-ordinate and co-operate with other regulators in a particular policy domain or even regarding a specific regulatory enterprise (second tier). Farther on, a brief overview is provided of various incentives for co-operation on both levels. However, the reasons for co-operating on both levels cannot be strictly separated, and both levels of co-operation should be seen as complementing each other.

Reasons for first tier co-operation

A number of reasons for the first form of co-operation have already been given above, when addressing the emergence of TPR in general. In addition, the following incentives behind first tier co-operation in TPR were observed:

Efficiency, inter-firm co-ordination and competition, collusion

Some forms of private regulatory co-operation are dictated by the need to enhance inter-firm co-ordination for pro-competitive purposes. This is typically the example of patent pools. TPR here is instrumental for increasing market shares (see Lerner and Tirole, 2004, p. 691; Merges, 1996; Gallini, 2011, p. 5). Other forms of co-operative arrangements can hide strategic anti-competitive purposes, as is the case with collective boycott schemes and cartels. Also, existing (and often concealed) private agreements between content providers and internet service providers for the purpose of disabling copyright infringement in cyberspace, often go beyond what public regulation would entail, especially in terms of copyright scope and respect for user privacy and network neutrality (e.g., MoU against Copyright Infringement).

Complementing or pre-empting public regulation

Some private regulatory co-operation initiatives emerge as a complement to existing public policy, to facilitate implementation and compliance with existing public regulation, either at state or transnational level. The former are called vertical and the latter horizontal complementarity (Cafaggi, 2006, p. 357, and 2011a). Institutional complementarity between private and public regulators emerges in various forms, at standard-setting level, monitoring and enforcement (Cafaggi, 2006, 2011b). Such complementarity can translate into formal agreements or into informal co-ordination. The formal emerges in the field of technical standardisation, i.e. of the "new approach" to standardisation in force in the EU since Directive 98/34 of 22 June 1998 (Cafaggi, 2006, 2011a). The informal institutional complementarity characterises the field of advertising where informal co-regulation between public and private occurs on both the vertical and the horizontal dimensions. At international level, the European Advertising Standards Alliance and the European Commission co-operate with each other; whereas national self-regulatory organisations co-operate with their national governments to enforce deceptive and misleading advertising rules.[47]

A third case, where not only complementarity but also coherence is an important factor, is the co-operation between ISO, the UN Global Compact, the OECD and ILO in the 26000 drafting process.[48] In the field of food safety, to some extent GlobalG.A.P. can be seen as complementing existing public regulation in the food chain. As an example, Sorsa (2011) states that "without GlobalG.A.P., farmers and exporters in DCs would have to come to grips with complicated EU regulations on food safety and those of the member states as well", and that in this sense, "private food schemes help to reduce the transaction costs by systematically providing information about European regulations on food safety systematically available and practically achievable". At the same time, some private governance schemes were created with a clear objective to pre-empt and avoid public regulation (product labelling and corporate accounting provide illustration) both at state and international level.

Enhancing legitimacy and effectiveness

Many forms of co-operation arise to increase legitimacy of both the regulatory process and output in form of standards, guidelines, etc. As one would expect, not only the private seeks public support to enhance its legitimacy and expand the reach of its regulatory influence but the public, often deploying soft law, can also borrow legitimacy from private schemes. In many IGOs, recent regulatory instruments express recognition of privately drafted codes and guidelines. Similarly, for effectiveness, forms of regulatory co-operation emerge because individual schemes do not reach a satisfactory level of compliance. Commercial contracts are used to ensure compliance with human rights policies. To pursue a higher level of responsiveness and consistency, co-operation among regulators is often combined with co-operation between regulators and regulated entities.[49]

Controlling the value chain

Other forms of co-operation schemes were created to reinforce control of the global supply chains, especially in the case of large retailers relying on local suppliers in countries with a very weak rule of law. In the area of food safety, the adoption of the supply chain approach by public regulation earlier in Europe and later in the US has delegated to and provided large MNEs with the regulatory power and responsibility to control safety along the chain (Gereffi, et al., 2005; Cafaggi, 2010, 2012b). In the area of CSR, codes are often implemented via contracting as illustrated by the case of private security services (see Francioni and Ronzitti, 2011; Dickinson, 2011). In the first period, competition rather than co-operation was the predominant feature. The excess of regulatory fragmentation led to overlapping schemes without real additional benefits for the final regulatory beneficiaries, the consumers, and even for the regulators themselves. This excess of private regulatory competition is leading to stronger or weaker forms of co-operation (e.g. the GFSI benchmarking scheme, corporate social responsibility and food safety).

Overcoming missing competencies

According to the Abbott and Snidal expertise, independence, representativeness, and operational capacity depend, to varying degrees, on the regulatory stage (standard setting, monitoring or enforcement), necessary competences in the regulatory process. The different actors in the process, however, are not equally strong in these categories. States might lack business expertise, whereas businesses may not be sufficiently representative or independent. NGOs on the other hand often lack operational capacity, but have solid expertise or independence (Abbott and Snidal, 2010). Often the reason for co-operation is to bundle complementary competencies.[50] In particular public/private co-operation might stem from the necessity to build on different competencies that transnational regulators have developed over time. Therefore, in the field of environment, data protection, accounting, financial markets, or internet regulation, just to name a few, co-operation built on actors with different competencies takes place.[51]

Reasons for second tier co-operation

Second tier co-operation, as outlined above, often takes place for similar reasons as within the first tier. Yet, there are some specificities worth examining:

Proliferation and fragmentation

Private schemes are often developed in a disorderly fashion. Sometimes the development occurs as a spontaneous phenomenon, other times it is promoted by public actors willing to refer to single schemes when regulating cross-border matters, as was mostly the case for the corporate disclosure of environmental, social and governance (ESG) reporting. Proliferation translates into transnational regulatory competition that drives evolution and innovation but, at the same time, might increase fragmentation. Regulatory co-operation among private regulatory scheme owners may represent, as outlined earlier, a potential response to the shortcomings of proliferation.

Conflict avoidance and creation of coherency

Competition may occur among regimes representing similar or identical interests. Regulatory schemes may also compete when they advance divergent interests and pursue conflicting objectives. For example, agricultural regimes may conflict with environmental ones, and trade and environmental regimes may be at odds with each other. Various forms of regulatory co-operation address conflict prevention or conflict resolution. In contrast to the public domain, where conflicts between international regimes and between regulators have received significant attention in academic literature (see ILC, 2006), conflicts in private regulation have hardly been examined, even if the practice shows a voluminous array of new instruments to deal with incompatibilities (for a first mapping, see Cafaggi, 2011c).

Most co-operative forms include some conflict preventing elements, since they usually, at least, slightly curb the risk of conflict. Entities create an environment where they have the opportunity to exchange information and to collaborate and thereby address emerging conflicts at a very early stage.[52] Benchmarking and peer review characterised many regimes. Furthermore, they often provide for a harmonisation of their regulatory approaches which also helps avoiding conflicts at an *ex ante* stage.[53]

Memoranda of Understanding have been used for conflict avoidance purposes. The MoU between ILO/ISO related to ISO 26000 offers an example of a conflict resolution mechanism. Here several provisions stipulate that ILO is the competent organisation in the field of labour regulation and that its standards will prevail in the event of a conflict: "International labour standards adopted by the ILO will take priority in any case of conflict in the context of development, and of any promotion, support, evaluation and approval, or periodic review of any ISO International Standard in the field of SR, as well as in any case of conflict involving ILO issues with any private initiative with which ISO may collaborate in the context of Standards".[54] Interestingly, this provision does not only foresee priority of ILO standards in relation to ISO 26000 but it also obliges the ISO to prefer ILO standards in relation to any other private initiative it might collaborate with. In this way, it sets a priority for the entire system where ILO standards will prevail over any other standards in the event of a conflict (one potential exception are other public standards of course).

Not all approaches about conflicting regimes use hierarchy. Usually, agreements contain provisions that provide for adaptation or compatibility of standards whereby they reduce the potential for future conflicts. For example, the alliance between GRI and the GC provides for integration of "Global Compact issue areas and principles centrally in the GRI

Guidelines". It is further stated that "[t]he Global Compact will, with the support of GRI, develop guidance on the use of GRI as the recommended reporting language" (United Nations Global Compact, 2010). In the case of the MoU between FSC/LEI an analysis is provided for to assess the "compatibility of the LEI standards as national FSC Forest Management standards". Here the plan is to integrate and adapt one standard (LEI) into the other organisation's system (FSC).

Reducing transaction costs through standardisation

There is an increasing "delegation" both from public and from private regulators to technical standard setters. This delegation constitutes a form of transnational regulatory co-operation between IGOs, private regulators and technical standard setters. But there is also transnational co-operation among technical regulators. As a consequence, the degree and quality of co-operation between technical standard setters and other regulators has increased.

Technical standard-setting bodies may co-operate with each other, for instance to ensure inter-operability. As an example, ISO and other technical standard setters engage into co-operation with more specialised technical regulators when necessary;[55] or with other regionally positioned organisations in order to avoid duplication of effort.[56] In the area of social responsibility, inter-operability seems to be of importance as well. When ISO 26000 was created, considerable emphasis was put on ensuring the inter-operability of this standard with already existing regulation in the area. Apart from the participation of a great number of organisations in the standard-setting process, ISO engaged in further co-operation on implementation. For instance, it signed an MoU with the Global Reporting Initiative (GRI), to ensure that businesses can better implement ISO 26000 into their sustainability reporting within the GRI framework.[57]

Even though a private regulator might have the ability, as well as the competence, to conduct a particular regulatory activity, it might be more efficient to co-operate with one of several other regulators. For example, if standards already exist, the costly process of drafting a new one could be saved. Moreover, regulatees would not have to adapt their activities (e.g. reporting) to yet another regulatory regime but could comply with the various compatible standards without additional burdens. Hence, for greater efficiency, there could be two incentives for regulators to co-operate with each other. First, to streamline and decrease the efforts and costs of the organisation itself by merging capacities in the regulatory process; and secondly, to decrease the effort and costs of the regulatees.

Enhancing quality and effectiveness

Quality and Effectiveness can be important incentives for private regulatory co-operation at various stages of the regulatory process. At the standard-setting stage, co-operative agreements, in particular those dealing with technical standard setting, put significant emphasis on increasing effectiveness through a common approach which does not necessarily imply uniform standards.[58] Co-operation allows these organisations to distribute the different tasks necessary in technical standardisation processes and to avoid duplication of efforts. The goal of a large number of co-operative projects is to streamline or harmonise existing standards to avoid duplicating impacts.[59] Global technical standards can be harmonised through various co-operative strategies. As examined above, mutual recognition is increasingly used to ensure that the outcome of monitoring compliance can trigger consequences without conflicting views. This problem has arisen in many certification regimes that were monitoring compliance with the same standards. However, co-operative approaches do not automatically increase effectiveness. The inclusion of a high number of stakeholders, the need to provide assistance, such as translations and other measures, can also decelerate a standard-setting process significantly and thus have an impact on its effectiveness.[60]

Improving effectiveness of compliance mechanisms

Private regulation is often (not always) on a voluntary basis. Compliance by the regulated is based on considerations such as cost-efficiency, self-interest, and reputational aspects. Pressure to comply may also originate from other private actors, such as investors or insurers, or from the public side, governments or national regulatory agencies. Incentives to comply are partly dependent on institutional design and partly driven by external environmental factors. Compliance may depend on the global regimes' flexibility and capacity to adapt to local circumstances. Often, agreements only contain provisions that deal with the implementation of private regulatory standards at local level.[61] Less common are provisions that explicitly stipulate co-operation regarding the compliance mechanism or even enforcement. An interesting exception is ISO 26000, which has provided a harmonised standard on social responsibility to help organisations (in particular businesses) to comply with these obligations. Remarkable in this context is Article 2.2.2 from the MoU between ILO/ISO, which stipulates that ISO's "activities … [will] [c]omplement the role of governments in ensuring compliance with international labour standards". Hence, through the co-operation with the industry-driven ISO, ILO also supports the efforts to ensure compliance with its own standards.

Signalling and enhancing legitimacy

Some regulatory co-operation schemes respond to signals sent by the market or by civil society (e.g., sustainability reporting, ethical trading, environmental protection and business "greenwashing", respect of labour rights, child labour etc.).[62] These include schemes that emerge due to the need to govern common resources sustainably (e.g. in the case of the FSC and the MSC).[63] A specific type of signalling occurs when a private regulator seeks to increase the legitimacy of its activities. For instance, when ISO decided to expand its scope into the area of social responsibility it faced a number of obstacles, in particular concerning question of representation. Given its traditional tendency to have mostly industrialised countries and businesses dominating, ISO sought to involve other stakeholders by creating a system including six groups: Consumer, government, industry, labour, NGO, and others. Part of this more inclusive process also resulted from MoUs with the relevant international organisations in the area of social standard setting (see Diller, 2012, p. 481). Here, the need to ensure that the regulatory process is legitimate and its final outcome was an important reason to engage into co-operation.

Broadening the scope of regulatory activities

For regulators it might become necessary to develop and broaden their scope of activities into areas where they have not yet been active. If this happens regulators might not have the capacities to regulate in the new field. There are many examples of this: In many agricultural standards it has become a practice to include environmental and community protection standards.[64] When the IOC integrated sports as the third pillar of the Olympic Movement, it relied to a large degree on co-operation with environmental organisations (in particular UNEP) to assist it in coping with the additional requirements of the field.[65] ISO expanded from a purely technical standard setting into areas with a greater public policy impact, such as environmental protection or social responsibility. As various other organisations, and particularly IGOs, such as the UN, the OECD and the ILO, were already working in this field, co-operation became a way to foster activities in the new area and to ensure that they fall within the existing framework.

Achieving competitive advantage

Co-operation can provide competitive advantage for the parties engaged in the process at the expense of outsiders. They might be able to set more capacities free, to attract more regulated, to reach more beneficiaries, and thus potentially gain higher legitimacy and greater effectiveness. The new situation could trigger additional growth in terms of regulatory impact. The drawback is that the additional regulatory power could also be exploited and the positive effects of regulatory competition be reduced. Second, standards could become diluted if too many interests have to be considered in the co-operative project. Concrete examples of co-operation to achieve competitive advantages include, in the forestry sector, the collaboration between FSC and LEI, which was reportedly triggered by the need to challenge competing organisations such as the Programme for Endorsement of Forest Certification (PEFC).[66]

Assessment

Today, TPR is an important source of rules and standards for companies and civil society organisations operating across the globe, with a myriad of standards and principles being developed at sectoral and regional levels. In key cross-cutting areas such as sustainability reporting or the food chain, fragmentation has generated a need for consolidation, which was initially examined by private meta-regulators such as ISEAL, and is now increasingly endorsed by international organisations such as FAO. In addition to this new trend, indicators are being used to signal the performance of individual regulators at globally agreed thresholds – a tendency that is still resisted by many industry players, who believe global thresholds and performance scores do not do justice to the diversity of conditions in which businesses operate in the various regions of the world. These developments seem likely to bring more legitimacy, quality and effectiveness to TPR, although there are still sceptics that argue they are likely to permit more sophisticated forms of "greenwashing" by large enterprises, and potential new barriers to market access for smallholders and SMEs. In response to these criticisms, there is growing calls for transparency and accountability of international organisations that support the adoption of new assessment frameworks, something that would probably contribute to making the efforts undertaken under their aegis reliable.

Against this background, there seems to be an emerging need for third parties to evaluate TPR schemes, and especially by public policy makers. Evaluation concerns regulatory performance and can be carried out within benchmarking schemes or in relation to individual regulators. One of the rationales for this evaluation lies in the possible difference between the private incentives and goals of TPR schemes, on the one hand; and public policy goals, on the other hand. This difference can be identified by reference to "effectiveness". Public policy makers might decide to delegate the solution of a given policy problem to TPR whenever this represents the most "effective" way to achieve that goal, for example, to increase compliance. In this respect, private regulation becomes simply one of many alternative policy options. However, the notion of effectiveness used here (defined as the extent to which private regulatory schemes achieve socially optimal or desirable outcomes[67]) might well differ from the one sought by private actors involved in private regulatory schemes. Occasionally, these two notions can coincide; but in many instances they may diverge – hence the scepticism of many social scientists and policy makers when it comes to evaluating private regulation.

Cases in which "private" and "social" effectiveness diverge can be of different types: collective action in private regulatory bodies can aim at socially sub-optimal outcomes (as in the case of cartels); the TPR scheme might generate negative externalities (e.g., GlobalG.A.P. has been accused of creating barriers to trade for developing countries); the scope of the private regulatory scheme might be narrower than the impacts generated by its participants' activity (e.g. safety-oriented certification schemes might generate unintended environmental consequences); or there might be cases in which the private regulatory scheme is aimed at achieving socially desirable outcomes, but either adverse selection problems or lack of monitoring and compliance leads to the emergence of socially undesirable outcomes.[68] Recent problems experienced by co-regulatory schemes at EU level (notably, in payments and in data protection) have been overshadowed by the growing "horse meat" scandal, which now extends to some of the largest producers of beef in the world, and is attributed to a joint failure of public regulation and private regulatory schemes in charge of auditing and inspecting outlets, thus complementing public regulation in the enforcement phase.

This potential misalignment between private benefits and social welfare can be dealt with by designing appropriate indicators that would make *ex ante* clear when regulatory choices pursue one or the other strategy. This could lead to the use of regulatory governance indicators, indicators related to regulatory objectives, performance indicators, etc. In addition, it would be important to note whether a given TPR scheme is likely to maintain its virtuous features over time. As a matter of fact, besides these "genetic problems", a number of other effects can undermine the alignment between private benefits and social goals during the life of a private regulatory body. These are briefly listed below:

- *Lock-in effects and collective action problems* can occur when members remain locked into sub-optimal agreements and "focal points", with no incentive to change;
- *Path dependency, status quo bias, anchoring and framing effects* may lead to shifting focus towards measurable and immediate benefits rather than long-term social welfare;

- *Hard-to-detect changes* over time might be induced by the prevalence of some interests over others during the life of the private regulatory body (e.g. MSC, see below);
- *Divergence of interests* between the regulators and the regulated, which lead the former to prefer short-term actions that maximise their likelihood of being re-appointed;
- *Self-indulgence in the evaluation* of private regulatory bodies, when governance arrangements entail self-evaluation, or lack of legitimacy of third parties in charge of evaluation.

All these problems deserve careful scrutiny before one can actually conclude that a given policy issue is a good candidate for efficient and socially effective private regulation.[69] What is still missing is a comprehensive theoretical framework for assessing TPR from the standpoint of public policy, during *ex ante* impact assessments or *ex post* sectoral screenings. In more detail, many policymakers around the world have not clarified under what conditions a private scheme could be a suitable, reliable way of achieving publicly recognised goals.

Focusing more on the assessment of TPR and national private regulation would bring about three important changes. First, explicit endorsement of existing schemes by national regulators would contribute to their legitimacy and their in-depth scrutiny from the standpoint of public policy. Second, it would force public regulators to develop comparative methodologies to select cases in which private regulation is likely to perform better than public regulation, or, in the light of institutional complementarity, which private scheme best fits the public regulation framework. Third, it might lead to better guidance for TPR schemes with regard to what governance arrangements, procedural requirements, regulatory tools and enforcement mechanisms they should adopt in order to meet minimum reliability thresholds from the perspective of society at large.

As most of the private schemes complement public regulation, the assessment exercise should focus on how to improve co-ordination and ensure cost-effective co-operation while pursuing public interest objectives. But the importance of self-assessment goes beyond monitoring the pursuit of public interest. In many private schemes, monitoring via benchmarking and peer review aimed at defining best practices and increasing mutual learning can be seen. These frameworks have been used by public domestic and international actors to evaluate the legitimacy and effectiveness of TPR.

Private governance schemes in the light of public regulation can be evaluated at various phases of the policy cycle. When self- or co-regulatory schemes are proposed in response to action by public policy makers, of course they will form part of the available regulatory options to be compared in an *ex ante* RIA. However, this is not always the case: many private governance schemes are created independently of public regulation, and this frequently leads commentators to exclude the possibility that a systematic monitoring of private governance by public regulators could ever take place. At the same time, there might be cases in which scrutiny by a regional or national policymaker is not very helpful, due to the global nature of private governance arrangements.

Both arguments, however, are not conclusive. Monitoring does not coincide with oversight. Incentives to monitor existing private schemes may serve the purpose of deciding if and what kind of public intervention is needed. In more detail, the evolution of the smart regulation agenda, especially in the EU, increasingly points at "closing the policy cycle", and thus at constantly monitoring the effects of existing regulatory schemes – whether public or private – through the use of indicators and a sequential, logically consistent use of *ex ante*, interim and *ex post* evaluation (European Commission, 2010b). Moreover, the fact that private governance arrangements tend to be global certainly implies that the best possible response would, in many cases, be an appraisal by international bodies or through public regulatory co-operation. However, nothing prevents a national or regional policymaker from assessing whether certain international rules are sufficient or desirable with respect to its own public policy goals.

That said, Cafaggi and Renda (2012) develop an evaluation framework along a number of sequential steps. Initially, the evaluation should consider the origin and type of TPR by identifying the rationales for creating the scheme (if already existing) and the phases of the policy cycle covered by TPR (agenda-setting, rule formulation, standards, implementation, monitoring, enforcement, etc.). Second, the evaluation of TPR should hinge on whether the governance of the TPR scheme can guarantee sufficient alignment between private benefits and social welfare: this might include the use of indicators such as participation, materiality, completeness, diversity of funding, specific governance arrangements, internal use of indicators and existence of self-evaluation or external evaluation arrangements. An example of a paper that uses indicators to assess the existence of such preconditions is Fuchs and Kalfagianni (2010). Third, evaluators should assess how the arrangements identified in Step 2 possibly affect the quality legitimacy and actual enforcement of the TPR scheme at hand (Cafaggi and Renda, 2012).

Once these phases have been completed, evaluators should consider effectiveness through three types of indicators: *i) activity and governance indicators*: i.e. indicators that correlate governance features and regulatory processes and outcomes of the private governance scheme; *ii) compliance indicators,* i.e. indicators used as means of reporting and signalling compliance with organisational goals helping to verify consistency between means and goals; and *iii) impact indicators,* which include criteria and indicators used to evaluate the performance of private regulatory schemes and their distributional impact on different constituencies; An example is whether the expected distributional consequences have occurred or wealth transfers are needed to correct unexpected effects. These latter "meta-indicators", in particular, are useful for public policymakers to understand whether private regulators evaluate themselves on the basis of "private" effectiveness.

Once these steps have been completed, the most appropriate tools that will host the evaluation will be an *ex ante* analysis, an interim or *ex post* evaluation. A number of additional filters could also be applied to private governance. The precise mix of specific methodologies that could be used in this respect is beyond the scope of this paper and could be the focus of additional work. A preliminary list may include tools that would support competition assessment, sustainability impact assessment, crime-proofing, testing for fundamental rights, specific risk assessment, sectoral competitiveness proofing and a policy coherence test.

Implementing dedicated guidance on how to assess private governance schemes would lead TPR to obtain full citizenship in the area of international regulatory co-operation (IRC). In this respect, TPR can prove decisive in contributing to IRC, by achieving a multi-stakeholder "deepening" of inter-governmental commitments, thus transforming them into concrete ways to achieve progress towards politically agreed goals on a global scale. Areas such as global commons, protecting human rights and, overall, achieving sustainability in global economic activities are perfect candidates for testing the reliability and effectiveness of public-private interplay in international regulatory co-operation.

Conclusion

The case study shows that forms of transnational co-operation involving private regulators are growing with new instruments constantly introduced in the global regulatory space. They take place within the private sphere between private regulators to reduce fragmentation, to prevent conflicts and mitigate uncertainty, to ensure the co-ordination of different policy objectives, to increase regulatory capacities in areas lacking strong regulatory frameworks or to govern the effects of private regulatory competition. Even more frequently, transnational regulatory co-operation emerges between public and private actors on standard setting, monitoring and enforcement. These forms of co-operation are quite common in relation to regulatory implementation, often decentralised at the local level, to monitoring, due to lack of resources on the public side, and to enforcement, given the effectiveness of some private dispute resolution mechanisms in relation to domestic decentralised enforcement.

Following this trend, new co-operative instruments, whose legal status is still undetermined have been developed. Informal co-operation via networks, roundtables, non-binding agreements and MOUs has proven to be quite effective in setting forth processes implemented by each organisation both on the public and the private side. Formal instruments like protocols, agreements, associations and partnerships have been created to sustain regulatory co-operation between public and private actors. There is no clear evidence that informal performs better than formal or *vice versa*. Their effectiveness is context dependent and varies according the institutional context.

The legal status of the public counterparts, whether treaty-based IGOs or transnational networks may influence both the instruments' choices and the effects of co-operative agreements with private regulators. Treaty-based IGOs have powers (limited) to engage with private actors and to delegate functions. Trans-regulatory networks often lack that formal power thereby resorting to informal arrangements. Different modes of co-operation can be found depending on whether private actors interact with international organisations or with trans-governmental networks. On the private side also, the status and identity of the individual transnational private regulator or the federation or association may have strong impact on both instruments' choices and effects. Regulatory co-operation with technical standard-setting bodies like ISO or private meta-regulators like GFSI or ISEAL differ from co-operation with individual enterprises or trade associations.

The co-operation between public and private actors can occur across different stages of the regulatory process. There are instances when the private defines the rules and the public enforces them or when the public sets the rules and the private monitors their compliance. There are examples of international soft law instruments whose compliance is monitored by private actors and the enforcement is ensured by a combination of domestic judicial, administrative enforcement and by private dispute resolution mechanisms. In other instances transnational private standard setting is endorsed at domestic European or national level by legislation and publicly enforced. This is often the channel through which global rules enter national legal orders. While these forms of vertical regulatory co-operation enhance effectiveness they require new instruments to ensure co-ordination between transnational actors and domestic legislatures and ensure legal certainty.

Therefore, the case study distinguishes transnational regulatory co-operation in relation to each stage of regulatory process: standard setting, monitoring, and enforcement. For each type a distinction can be made between vertical co-operation (i.e. a transnational body co-operates with one or multiple domestic/local regulators) and horizontal co-operation (i.e. two or more transnational bodies co-operate). A second structural dimension influencing the intensity of interactions and the dynamics of co-operation suggests differentiation between bilateral and multilateral co-operation, the latter encompassing either a web of bilateral linked agreements, or, less frequently, multiparty agreements.

The choice to co-operate and the selection of the appropriate legal instruments supporting TPR are not yet driven by a specific and well-defined set of principles – such as those that have developed for domestic regulatory policy culminating with the OECD Council Recommendation on regulatory Policy and Governance. Choices within the public domain are often made on an ad hoc basis, sometimes without a technical analysis concerning the consequences of selecting potential regulatory alternatives. Transnational regulatory co-operation should be decided on the basis of an impact analysis based on an informed selection of alternative strategies and instruments, encompassing the entire co-operative process: its inception, implementation and termination. The regulatory cycle should be broken down into standard setting, monitoring and enforcement. This evaluation should incorporate specialised indicators to analyse the effectiveness of co-operative transnational regulation.

Going further, there is a need for general guidelines that would provide guidance on when and according to which modalities the new forms of co-operation between public and private organisations are beneficial and can usefully complement existing regulatory mechanisms. Transnational co-operation can increase legitimacy of both public and private actors by making processes and effects transparent while promoting the inclusion and participation of regulated entities and regulatory beneficiaries. Nonetheless, a general cross-sectoral framework providing guidance and criteria in the identification and selection of beneficial TPR would help to avoid conflicts among transnational regimes and would increase legal certainty.

Further research is needed to define a proper taxonomy of TPR (encompassing both forms of co-operation between private regulators and between private regulators and public international institutions) based on the correlation between the status of the co-operating actors, the quality of the regulatory instruments governing co-operation and the objectives of co-operation. Only on the basis of this research can a proper institutional framework be usefully provided and serve the purpose of supporting effective and accountable public/private co-operation.

Notes

1. The ISEAL Alliance is the global association for sustainability standards. It was founded in 2002. For further information see www.isealalliance.org/.

2. See generally: Gereffi, et al. (2005), "The Governance of Global Value Chains", No. 1, *Review of International Political Economy*, p. 78 ff. In the area of food safety a great number of retailer driven organisations and incentives have been created in order to ensure the safety of food products throughout the entire chain, see FAO, Food Safety Certification, 2006; Gereffi, A global value chain approach to Food Safety and Quality Standards, Paper prepared for the Global Health Diplomacy for Chronic Disease Prevention Working Paper Series (February 2009); Cafaggi, Private Regulation, Supply Chain and Contractual Networks: the Case of Food Safety, 2010 available on ssrn. Regarding social and environmental regulation numerous retailers have become active in implementing rules that they make mandatory for their suppliers, thereby ensuring minimum standards to be complied with. See, e.g. in the case of Wal-Mart, Standards for Suppliers, www.walmartstores.com/AboutUs/279.aspx.

3. ICANN is a non-profit private organisation, created in 1998, responsible for the co-ordination of the global Internet's systems of unique identifiers.

4. For more information see: http://worldstandardscooperation.org/.

5. See ISO international standards and "private standards", www.ISO.org.

6. Conservation agreements are usually offered by or in co-operation with governments and allow certain benefits to property owners in return for them protecting the environmental and/or cultural features on their terrain (see e.g. www.environment.nsw.gov.au/cpp/ConservationAgreements.htm), other forms of regulation by contract, e.g. supplier agreements, where (usually large) retailers stipulate social and environmental standards that suppliers have to comply with. See, e.g. in the case of Wal-Mart, Standards for Suppliers, cit. above.

7. In many areas of private regulation (such as food safety, fair trade, CSR and environmental protection) public regulation is incorporated into the respective standards. E.g. Fairtrade International states: "When setting the Fairtrade Standards, Fairtrade International (FLO) follows certain internationally recognised standards and conventions, particularly those of the International Labour Organization (ILO)." (quote taken from the Standards for Small Producer Organizations, www.fairtrade.net/fileadmin/user_upload/content/2011-12-27_spo_en_final.pdf). Principle 1 of the FSC 10 Principles states: "Compliance with laws and FSC Principles – to comply with all laws, regulations, treaties, conventions and agreements, together with all FSC Principles and Criteria", www.fsc.org/the-ten-principles.103.htm.

8. The best example in this regard are sport organisations, such as the IOC, FIFA etc., which have a rather undisputed authority within their area of expertise and provide for an almost complete set of institutions including a "judicial" (arbitration) branch (Court of Arbitration for Sport). However, there are also a number of examples where such regimes "clashed" with national or regional public law provisions. In this regard see Court Judgment of 15 December 1995. Union royale belge des sociétés de football association ASBL v Jean-Marc Bosman etc., Case C-415/93.

9. An example for the first case is the Global Reporting Initiative (GRI), which provides for a framework under which organisations (companies and others) can publish their sustainability records. Its reporting framework consists of a general part and of sector- and country-specific guidelines: www.globalreporting.org/reporting/Pages/default.aspx. The most common example is the ISEAL Alliance, which works out codes of good practice for standard setting and implementing, see: www.isealalliance.org/about-us.

10. ISEAL Alliance members include both the FSC and the MSC. Other founding members include Fairtrade Labelling Organizations International (FLO), the International Federation of Organic Agriculture Movements (IFOAM), the International Organic Accreditation Service

(IOAS); the Marine Aquarium Council (MAC); the Rainforest Alliance; Social Accountability International (SAI); Social Accountability Accreditation Services (SAAS), the Union for Ethical BioTrade (UEBT) and UTZ Certified. To become a full member of the ISEAL Alliance, members have to demonstrate full compliance with ISEAL Codes of Good Practice and other applicable ISO Guides (e.g. ISO17011 for Accreditation Bodies). Organisations interested in membership have to successfully complete a pre-assessment. ISEAL has historically relied upon three sources of funding – governmental agencies, foundations and membership fees.

11. For more information see ISEAL Alliance (n.a.), "Our Codes of Good Practice", www.isealalliance.org/our-work/codes-of-good-practice.

12. For example, take the case of ISO, which is constituted by national standard-setting bodies; a number of which are public entities. Furthermore, technical standards are incorporated into public international documents such as the Technical Barriers to Trade Agreement (Article 2.4).

13. As a classic example, one can name here the UN Global Compact, which is considered as a platform provided by the UN for business to align its operations with common goals in the area of environment, human rights, labour standards and anti-corruption efforts. See www.unglobalcompact.org.

14. For more information see www.unglobalcompact.org/AboutTheGC/index.html and Waddell (2011).

15. As an example, see the Memorandum of Understanding between Internet Platforms and Rights Owners. Most signatories are business entities which are predominantly concerned with common business activities, such as creating profits and which only engage into regulatory activities on a secondary basis.

16. As an example, see the Agreement between the WWF and GlobalG.A.P. on Aquaculture Standards.

17. International regulatory organisations refer to formally established organisations (e.g. as an association, a foundation or a corporation) whose main purpose is regulation, in contrast to other entities such as business or NGOs, whose main purpose is profit making or political advocacy. Examples of regulatory organisations are the ISEAL Alliance, as an organisation whose membership is mainly constituted by other regulatory organisations such as the FSC (see: www.isealalliance.org), but also other organisations such as GlobalG.AP., whose members are food producers and suppliers and who joined in order to effectively regulate food safety (see: www.globalgap.org).

18. See in particular in the financial markets sector, where rating agencies work under a for-profit model but are nonetheless usually referred to as regulators.

19. See the FSC, which states on its homepage: "FSC is an independent, non-governmental, not-for-profit organisation …" www.fsc.org/about-fsc.html.

20. A small selection of examples: the FSC set up by different social and environmental groups as well as the industry, GlobalG.A.P. bringing together retailers and producers in order to set food safety standards, international sports associations consistingof their national members, ISO as an international standard setter consisting of national standard-setting bodies or, in financial regulation, the International Swaps and Derivatives Association which brings together banks and participants in the derivatives markets, etc. Regarding forms of public private co-operation, one can name the European Strategic Safety Initiative, which is more a network (or partnership) but which, as a privately structured initiative, includes EASA, other public regulators and the industry. For more information, see: http://easa.europa.eu/essi/.

21. For instance, the MoUs concluded between ISO and the different IGOs for the ISO 26000 standard-setting process might not be considered legally binding *stricto sensu* and they are certainly not enforceable by the courts. Yet, the risk of one of the parties withdrawing and the negative consequences that would have emerged for the whole process, created a strong incentive to adhere to the provisions within the respective agreements.

22. This is the case regarding a self-commitment agreement concluded by chemical companies using renewable as well as fossil raw materials as feedstock for their production. Here, a working group of the DG Enterprise and Industry played a facilitating role. Generally the Commission declares its interest in co-regulatory initiatives in the area of CSR, see European Commission, A renewed EU strategy 2011-14 for Corporate Social Responsibility, COM(2011) 681 final, 25 October, 2011. Also in the realm of internet regulation, see the example of the safer social networking principles as well as the Pan-European Video Game Classification Scheme that both took place through the facilitation of the European Commission.

23. See www.efrag.org/Front/n1-1102/EFRAG-and-the-FRC-issue-the-Feedback-Statement-on-the-Discussion-Paper--Improving-the-Financial-Reporting-of-Income-Tax.aspx.

24. See UNEP and EPA (2011), MoU between UNEP/EPA (US Environmental Protection Agency) or UNEP and the European Commission on environmental matters.

25. See FSC – LEI (2001), MoU between the FSC/LEI; or GlobalG.A.P./WWF (Aquaculture Standards); or ISO – CEN (1991), Vienna Agreement between ISO and CEN.

26. See MoU between Internet Platforms and Rights Owners.

27. Seethe agreements concluded between UNEP and different sports organisations.

28. SeeIPL and UNEP (2010), MoU between UNEP/IPL; or FSC – LEI (2001), MoU between FSC/LEI.

29. See MoU between ILO/ISO, OECD/ISO and GC/ISO regarding the creation of ISO 26000.

30. See MoU between OECD/GRI; UNEP/Sochi2014 which will terminate on 30 June, 2014.

31. An example of such a rather provisional agreement is the MoU on Working Arrangements between the Asian Development Bank and the UNEP, which first outlines the principle areas of partnership (Section C.) and then provides for exchange of information and consultation (Section D.). Similarly, the MoU between GC/EFQM also sets out the framework for a partnership, more detailed information added on how this will be concretely designed, and then includes a number of procedural provisions.

32. The MoU between IEC/ISO/ITU/UN-ECE contains procedural and substantive aspects: "In the context of open-edi, it is understood that ISO and IEC are responsible for the development of standards concerning security in edi transmission (…). UN/ECE and the participating International User Groups are invited to contribute to this work by providing input through liaison and through direct participation of their experts."

33. See Article 5 of the Agreement between ISO/TC 211 Geographic information/Geomatics and the Digital Geographic Information Working Group (DGIWG).

34. The MoUs in the context of ISO 26000 all stipulate for participation within the ISO framework (Article 5 MoU between ILO/ISO; Article 4 MoU between OECD/ISO; Article 4 MoU UNGC/ISO) as well as for additional co-operative activities.

35. Examples for the former are manifold, see only: Article 3 MoU between ILO/ISO; Article 5 MoU between OECD/ISO; MoU between Internet Platforms and Rights Owners, Memorandum of Understanding against Copyright Infringement (see www.copyrightinformation.org/sites/default/files/Momorandum%20of%20Understanding.pdf); within organisations information exchange usually takes place in form of "stakeholder input" regarding the different regulatory processes

conducted by the organisation. E.g. the FSC foresees such input regarding its standard-setting process: www.fsc.org/standard-setting.212.htm. Similar processes can be found within most regulatory organisations. ISEAL Alliance, an organisation providing codes for regulatory processes within a number of private environmental and social organisations requires such input as a criteria of good standard setting, see www.isealalliance.org/our-work/codes-of-good-practice/standard-setting-code.

36. See, e.g., MoU between UNEP/FELABAN; MoU between the FSC/LEI; MoU between the Indian Premier League, a Sub-committee of the Board of Control for Cricket in India (BCCI) (IPL) and the United Nations Environment Programme (UNEP) (see www.unep.org/sport_env/images/IPL_MoU.pdf).

37. Abbott and Snidal, The Governance Triangle, 44, at 46 use a five-step regulatory process: agenda setting, negotiation, implementation, monitoring and enforcement.

38. In the case of GlobalG.A.P. both stages are organised by the organisation itself. However, as with the other institutions, certification is conducted by independent bodies which are accredited to the system, see www.globalgap.org/cms/front_content.php?idcat=2. Accreditation and supervision of the certification bodies were outsourced by the FSC and the MSC to ASI (Accreditation Services International), which the FSC founded for this purpose, see www.accreditation-services.com/about/asi.

39. ISO is a good example of this case. It sees itself as a developer of standards and does not engage into monitoring or enforcement, see www.iso.org/iso/about.htm. Other examples are ISDA, or the International Accounting Standards Board (IASB) and the Financial Accounting Standards Board (FASB), to name just a few.

40. ISO standards are usually certified by independent certifiers. Accounting standards are enforced mainly through the public. See furthermore, Zimmermann, et al. (2008).

41. This for example is the case with ISO 26000, see www.iso.org/iso/iso_catalogue/management_and_leadership_standards/social_responsibility/sr_discovering_iso26000.htm#std-1. Furthermore, as an example, the Pan-European Video Game Classification Scheme sets standards with the involvement of the European Commission. However, to a large extent, it leaves the enforcement (in addition to its own filtering system) to national regulation. For more information see www.pegi.info/en/.

42. See Article 1.1 MoU between ILO/ISO; Article 1.1 MoU between OECD/ISO; Article 1.1 MoU between GC/ISO; MoU between LEI/FSC;

MoU between GRI/GC. See the Vienna Agreement between CEN and ISO.

43. As an example, see the Programme for the Endorsement of Forest Certification (PEFC), which can be taken as an umbrella organisation that endorses national standards of its national standard-setting members, see www.pefc.org/standards/national-standards. Furthermore, GlobalG.A.P. provides for a benchmarking standard, see GlobalG.A.P. Benchmarking, see www.globalgap.org/uk_en/what-we-do/the-gg-system/benchmarking/.

44. See Directive on the Promotion of the Use of Energy from Renewable Sources and Amending and Subsequently Repealing Directives 2001/77/EC and 2003/30/EC, 2009/28/EC, 23 April 2009. As well as Communication from the Commission on Voluntary Schemes and Default Values in the EU Biofuels and Bioliquids Sustainability Scheme, 2010/C/160/01 which provides: "Economic operators must show Member States that the sustainability criteria relating to greenhouse gas savings, land with high biodiversity value and land with high carbon stock have been met. They can do this in three ways:

    1. By providing the relevant national authority with data, in compliance with requirements that the Member State has laid down (a "national system");

    2. By using a "voluntary scheme" that the Commission has recognised for the purpose;

    3. In accordance with the terms of a bilateral or multilateral agreement concluded by the Union with third countries and which the Commission has recognised for the purpose."

45. See in particular Article 6 of the Regulation Laying down the Obligations of Operators who Place Timber and Timber Products on the Market, No. 995/2010, 20 October, 2010. See C. Overdevest and J. Zeitlin, (2012).

46. For more information see: www.efrag.org/Front/c1-262/efrag-Facts.aspx.

47. See www.easa-alliance.org.

48. See www.iso.org/iso/home/standards/iso26000.htm for more information.

49. A notable example is the co-regulatory scheme created by the European Commission in the field of payments, which aimed at the creation of a pan-European payment scheme (SEPA). A private institution – the European Payment Council – was created to strengthen co-operation and co-ordination between the regulated entities and the European Commission. See http://ec.europa.eu/internal_market/payments/sepa/ec_en.htm.

50. There are many examples of this and they can be found in various co-operative agreements. See the preamble of several agreements that list the different competences a regulator can provide, such as the MoU between ILO/ISO; the MoU between IPL/UNEP; the MoU between GC/ISO. Or also see specific provisions such as the following: "2.2 That any ISO activities and/or publications for the promotion, support, evaluation and approval of any published ISO International standard on SR, insofar as they implicate ILO issues, will: Facilitate greater awareness and wider observance of international labour standards in accordance with their object and purpose, and their interpretation by the competent bodies of ILO; Complement the role of governments in ensuring compliance with international labour standards". MoU between ILO/ISO.

51. In the accounting sector see the participation of the European Commission via the European Financial Reporting Advisory Group in the International Accounting Standards Board. With regard to financial markets, the Commission has the possibility of concluding soft law agreements with the International Organization of Securities Commission. Finally, in the case of internet regulation the Pan-European Video Game Classification Scheme was mainly responsible for the game classification. However, the European Commission was involved as an advisory and enforcement of the standards was mainly left to national regulators.

52. See ILO and ISO (2005), Articles 3 and 4, MoU between ILO/ISO and MoU between Internet Platforms and Rights Owners (2011).

53. See the MoU between FELABAN/UNEP or the MoU between FAO/ILO.

54. See in particular Article 2.3, ILO and ISO (2005), MoU between ILO/ISO.

55. See, e.g., IEC, ISO, ITU and UN-ECE (2000), MoU between IEC, ISO, ITU and UN-ECE; IULTCS and ISO (2005), MoU between IULTCS/ISO. With regard to CEN co-operation agreements , see: www.cen.eu/cen/AboutUs/CENnetwork/Relations/MoUs/Pages/default.aspx.

56. See ISO – CEN (1991), Vienna Agreement between ISO and CEN.

57. "The MoU is intended to leverage the activities of the two organisations related to reporting and benchmarking by business and on sustainable development by sharing information on ISO standards and GRI programmes, teaming up with other partners, participating in the development of new or revised documents, joint promotion and communication", www.iso.org/iso/pressrelease.htm?refid=Ref1460.

58. See MoU between IEC/ISO/ITU/UN-ECE; MoU between Internet Platforms and Rights Owners.

59. See the MoU between GRI/GC; furthermore see the MoU between LEI/FSC.

60. An example for a very inclusive process is the creation of ISO 26000:2010. See for a detailed description Diller (2012). On the different correlations between legitimacy and effectiveness see Cafaggi, New Foundations, p. 20.

61. MoU between LEI/FSC; Articles 3.1 and 3.5 of the MoU between IPL/UNEP, but see also the Alliance between GRI/GC, where it is stressed that the "GRI Sustainability Reporting Framework is a voluntary ESG reporting and stakeholder engagement and management tool and should not be viewed as a compliance framework", www.unglobalcompact.org/news/50-06-24-2010.

62. The term "green-washing" is used to refer to cases in which which green marketing is deceptively used to promote the perception that an organisation's aims and policies are environmentally friendly. American environmentalist J. Westerveld coined the term in 1986 in response to a hotel's efforts to encourage guests to help the environment by re-using towels. See, for an application to the oil market, Cherry and Sneirson, (2012), p. 133.

63. See, for an introduction to the governance of common resources, Van Waarden (2010). For applications to the MSC and FSC, see Gale and Haward (2011); and Curtin and Senden (2011).

64. See for instance the case of GlobalG.A.P which, apart from food safety, also covers environmental protection, workers' health, safety and welfare, as well as animal welfare. See GlobalG.A.P, Cultivating the Future of the Planet, www.globalgap.org/uk_en/what-we-do/globalg.a.p./.

65. See Sport and Environment Commission, www.olympic.org/sport-environment-commission.

66. See Frequently Asked Questions Memorandum of Understanding between FSC and LEI, www.fsc.org/fileadmin/web-data/public/document_center/news/press_releases/fsc-ss-2010-07-14-lei_fsc_mou_faq-en.pdf.

67. This is why effectiveness is normally linked to explicitly stated general, specific and operational objectives in the *ex ante* IAs of the European Commission. The specification of objectives has become much more common in Commission IAs over the past few years, and the Communication on smart regulation of October 2010 placed even more emphasis on the need to define "SMART" objectives in *ex ante* policy appraisal documents, so that achievement of those objectives can be monitored over time, including in *ex post* evaluation.

68. An example is provided by Lennox and Nash (2003) who describe the Responsible Care initiative launched by the Chemical Manufacturer's Association (CMA) in 1989 in response to growing public criticism of the industry: the fact that the CMA did not require third party review or certification of firm performance and did not adopt explicit sanctions for non-compliance led to a perverse situation in which participants in Responsible Care were more polluting on average than other chemical firms in the United States. Similarly, Morgenstern and Pizer (2007) in reviewing a number of voluntary programmes in the environmental field express concern on the self-selection of participants into those schemes.

69. For example, Ashby, et al. (2004) distinguished a number of voluntary regulatory schemes in the UK based on the different context in which they emerge, which in turn determines a different mode of strategic interaction between private players. Accordingly, they define the UK advertising Code as an Assurance Game, the UK Press as a Chicken game and the UK Life Insurance as a Prisoner's Dilemma.

Bibliography

Abbott K. W. and D. Snidal (2012), "Taking responsive regulation transnational: Strategies for international organizations", *Regulation and governance*, Wiley Publishing Asia Pty Limited.

Abbott, K. W. and D. Snidal (2010), "International Regulation without International Government: Improving IO Performance through Orchestration", Vol. 5, *Review of International Organizations*, p. 315.

Ashby, S., S. H. Chuah, and R. Hoffmann (2004), "Industry Self-Regulation: A Game-Theoretic Typology of Strategic Voluntary Compliance", *Working Paper*, Financial Services Authority, London.

Backer, L. C. (2007), "Economic Globalization and the Rise of Efficient Systems of Global Private Lawmaking: Wal-Mart as Global Legislator", Vol. 39, No. 4, *University of Connecticut Law Review*, 1739.

Bruszt, L. and G. A. McDermott (2009), "Transnational Integration Regimes as Development Programmes" in L. Bruszt, and R. Holzhacker (eds.), *The Transnationalization of Economies, States, and Civil Societies*, Springer, p. 23.

Büthe, T. and W. Mattli (2011), *The New Global Rulers: The Privatization of Regulation in the World Economy*, Princeton University Press.

Büthe, T. (2010), "Global Private Politics: A Research Agenda", Vol. 12, Issue 3, *Business and Politics*, p. 1.

Cafaggi, F. (forthcoming), Conflicts and Cooperation in Transnational Private Regulation, forthcoming, on file with the author.

Cafaggi, F. (2013), "The Regulatory Functions of Transnational Commercial Contracts. New Architectures", in Vol. 36, *Fordham Journal of International Law*, forthcoming.

Cafaggi, F. (2012), *The Enforcement of Transnational Regulation. Ensuring Compliance in a Global World*, Edward Elgar Publishers.

Cafaggi, F. (2011a), "New Foundations of Transnational Private Regulation", Vol. 38, *Journal of Law and Society*, EUI WP, p. 20.

Cafaggi, F. (2011b), "Private Regulation and European Private Law", in A.S. Hartkamp, et al. (eds.), *Towards a European Civil Code*, Chapter 5, Kluwer Law International, The Hague.

Cafaggi (2011c), "The Architecture of Transnational Private Regulation", *EUI Working Papers*, Law 2011/12, European University Institute, www.eui.eu.

Cafaggi (2010), "Private Regulation, Supply Chain and Contractual Networks: The Case of Food Safety". *EUI Working Papers*, RSCAS 2010/10, Robert Schuman Centre for Advanced Studies, Private Regulation Series-03, Florence.

Cafaggi, F. (2006), *Reframing Self-Regulation in European Private Law (Private Law in European Context),* Kluwer Law International.

Cafaggi, F. and A. Renda (2012), "Public and Private Regulation. Mapping the Labyrinth", *CEPS Working Document*, No. 370, Centre for European Policy Studies, Brussels.

Cashore B. and S. Bernstein (2004), "Non-State Global Governance: Is Forest Certification a Legitimate Alternative to a Global Forest Convention?", in J. Kirton and M. Trebilcock (eds.) *Hard Choices, Soft Law: Combining Trade, Environment, and Social Cohesion in Global Governance*, Ashgate Press, p. 33.

Cashore, B., G. Auld and D. Newsom (2003), "Forest Certification (Eco-labeling) Programs and their Policy-Making Authority: Explaining Divergence Among North American and European Case Studies", Vol. 5, Issue 3, *Journal of Forest Policy and Economics*, p. 225.

Cassese, S. (2005), "Global Standards for National Administrative Procedure", Law and Contemporary Problems", Vol. 68, p. 109.

Cassese, S. et al. (eds) (2012), *Global Administrative Law: The Casebook*, 3rd ed., Institute for IRPA – International Law and Justice.

Cherry, M. A. and J. F. Sneirson (2012), "Chevron, Greenwashing, and the Myth of 'Green Oil Companies'", Vol. 3, *Journal of Energy, Climate, and the Environment*, p. 133.

Coglianese C. (2011), "Measuring regulatory performance: Evaluating the impact of regulation and regulatory policy", OECD, Paris.

Coglianese C. and E. Mendelson (2010), "Meta and self regulation", in *The Oxford Handbook of Regulation*, Martin Cave, Robert Baldwin, Martin Lodge (eds.), *University of Pennsylvania Law School, Public Law Research Paper No. 12-11*; *University of Pennsylvania, Institute for Law and Economics Research Paper No. 12-06.*

Cooperative Agreement between ISO/TC 211, "Geographic Information/Geomatics and the Digital Geographic Information Working Group", www.isotc211.org/Agreements/Agreement_DGIWG.pdf.

Curtin, D. and L. Senden (2011), "Public Accountability of Transnational Private Regulation: Chimera or Reality?", Vol. 38, *Journal of Law and Society*, p. 163.

Davis, K. E. et al. (eds) (2012), *Governance by Indicators: Global Power Through Classification and Rankings*, Oxford University Press.

Dickinson, L. (2011), *Outsourcing War and Peace: Protecting Public Values in a World of Privatized Foreign Affairs*, Yale University Press.

Diller, J. (2012), "Private Standardization in Public International Lawmaking", Vol. 33, Issue 3, *Michigan Journal of International Law*, p. 481.

European Commission (2011a), "A renewed EU strategy 2011-14 for Corporate Social Responsibility", COM(2011) 681 final, 25 October, http://eur-lex.europa.eu/lexuriserv/lexuriserv.do?uri=com:2011:0681:fin:en:pdf.

European Commission (2011b), "Commission Staff Working Paper, Operational Guidance on Taking Account of Fundamental Rights in Commission Impact Assessments", SEC(2011) 567 final, 6 May, http://ec.europa.eu/governance/impact/key_docs/docs/sec_2011_0567_en.pdf.

European Commission (2010a), "Communication from the Commission on Voluntary Schemes and Default Values in the EU Biofuels and Bioliquids Sustainability Scheme", *Official Journal of the European* Union, 2010/C/ 160/01.

European Commission (2010b), "Smart Regulation in the EU", COM(2010)543, European Commission Communication, 8 October.

European Commission (2009), "Directive on the Promotion of the Use of Energy from Renewable Sources and Amending and Subsequently Repealing Directives 2001/77/EC and 2003/30/EC, 2009/28/EC", 23 April, http://eur-lex.europa.eu/LexUriServ/LexUriServ.do?uri=Oj:L:2009:140:0016:0062:en:pdf.

European Commission (2006), "External Trade: Handbook for Trade Sustainability Impact Assessment", http://trade.ec.europa.eu/doclib/docs/2006/march/tradoc_127974.pdf.

European Commission (2002), "Regulation on the Application of International Accounting Standards", No. 1606/2002/EC, 19 July.

Fairtrade International (2011), "Fairtrade Standard for Small Producer Organizations", www.fairtrade.net/fileadmin/user_upload/content/2011-12-27_spo_en_final.pdf.

FAO (2006), Food Safety Certification, FOA, ftp://ftp.fao.org/docrep/fao/008/ag067e/ag067e00.pdf.

Forest Stewardship Council, "The 10 Principles: Ten rules for responsible forest management", www.fsc.org/the-ten-principles.103.htm.

Francioni, F. and N. Ronzitti (eds) (2011), *War by Contract, Human Rights, Humanitarian Law, and Private Contractors*, Oxford University Press.

FSC – LEI (2001), "Memorandum of Understanding between the Forest Stewardship Council (FSC) and the Lembaga Ekolabel Indonesia (LEI)", www.fwi.or.id/sertifikasi/sertifikasi34.pdf.

Fuchs, D. and A. Kalfagianni, (2010), "The Causes and Consequences of Private Food Governance", Vol. 12, Issue 3, *Business and Politics*, Article 5.

Gale, F. and M. Haward (2011), *Global Commodity Governance: State Responses to Sustainable Forest and Fisheries Certification*, Palgrave Macmillan.

Gallini, N. T. (2011), "Private Agreements for Coordinating Patent Rights: The Case of Patent Pools", IEL paper in Comparative Analysis of Institutions, Economics and Law No. 5, *Institutions Economics and Law Working Paper Series*.

Gereffi, G. (2009), "A global value chain approach to Food Safety and Quality Standards", paper prepared for the *Global Health Diplomacy for Chronic Disease Prevention Working Paper Series*, February.

Gereffi, G, J. Humphrey and T. Sturgeon (2005), "The Governance of Global Value Chains", *Review of International Political Economy*, pp. 78-104.

Guttenstein, E. et al. (2010), "A Conceptional Framework for Progressing towards Sustainability in the Agriculture and Food Sector", Discussion Paper, FAO, ISEAL Alliance.

IEC, ISO, ITU and UN-ECE (2000), "Memorandum of Understanding between the International Electrotechnical Commission (IEC), the International Organization of Standardization (ISO), the International Telecommunication Union (ITU) and the United Nations Economic Commission for Europe (UN-ECE) Concerning Standardization in the Field of Electronic Business", www.itu.int/ITU-T/e-business/files/mou.pdf, 24 March.

ILC (2006), "Fragmentation of International Law: Difficulties Arising from the Diversification and Expansion of International Law", Report of the Study Group of the International Law Commission, finalised by Martti Koskenniemi, 58th Session, U.N. Doc. A/CN.4/L.682, 13 April.

ILO (1949), "Agreement between the International Labour Organisation and the Food and Agriculture Organisation of the United Nations", www.ilo.org/public/english/bureau/leg/agreements/fao.htm.

ILO and FAO (2004), "Memorandum of Understanding between the International Labour Organization (ILO) and the Food and Agriculture Organization of the United Nations (FAO)", www.fao-ilo.org/fileadmin/user_upload/fao_ilo/pdf/ILO_FAO_MoU_2004.pdf.

ILO and ISO (2005), "Memorandum of Understanding between the International Labour Organization (ILO) and International Organization for Standardization (ISO) in the Field of Social Responsibility", http://inni.pacinst.org/inni/corporate_social_responsibility/MoU%20ILO%20&%20ISO.pdf, 4 March.

IPL and UNEP (2010), "Memorandum of Understanding between the Indian Premier League, a Sub-committee of the Board of Control for Cricket in India (BCCI) (IPL) and the United Nations Environment Programme (UNEP)", www.unep.org/sport_env/images/IPL_MoU.pdf.

ISEAL Alliance (n.a.), "Our Codes of Good Practice", www.isealalliance.org/our-work/codes-of-good-practice.

ISO (2004), "ISO/IEC 17011:2004, Conformity assessment – General requirements for accreditation bodies accrediting conformity assessment bodies", Geneva.

ISO (2010), "ISO 26000:2010 Guidance on Social Responsibility", Geneva, www.iso.org/iso/catalogue_detail?csnumber=42546.

ISO-CEN (1991), "Agreement on Technical Co-operation between ISO and CEN (Vienna Agreement), 27 June, http://isotc.iso.org/livelink/livelink/fetch/2000/2122/3146825/4229629/4230450/4230458/01__Agreement_on_Technical_Cooperation_between_ISO_and_CEN__Vienna_Agreement_.pdf?nodeid=4230688&vernum=0.

IULTS and ISO (2005), "Memorandum of Understanding between the International Union of Leather Technologists and Chemists (IULTCS) and ISO on Cooperation in the Development of Standards Associated with the Testing of Tanned Leather and Tanning Products", www.iultcs.org/pdf/mou_iultcs.pdf, 20 December.

Judgment of the Court (1995), "Union royale belge des sociétés de football association ASBL v Jean-Marc Bosman etc., Case C-415/93" – Before the Court of Justice of the European Communicaties", December, www.biicl.org/files/1915_c-415-93.pdf.

Kingsbury, B. (2009), "The Concept of 'Law' in Global Administrative Law", Vol. 20, Issue 1, *European Journal of International Law*, pp. 23-57.

Kingsbury, B., N. Krisch and Stewart, R. B. (2005), "The Emergence of Global Administrative Law", Vol. 68, *Law and Contemporary Problems*, p. 15.

Lennox, M. and J. Nash (2003), "Self-Regulation and Adverse Selection: A Comparison Across Four Trade Association Programs", Vol. 12, Issue 6, *Business Strategy and Environment*, p. 343.

Lerner, J. and J. Tirole (2004), "Efficient Patent Pools", Vol. 94, Issue 3, *American Economic Review*, p. 691.

Marx, A. (2008), "Limits to Non-State Market Regulation: A Qualitative Comparative Analysis of the International Sport Footwear Industry and the Fair Labor Association", Vol. 2, Issue 2, *Regulation & Governance*, p. 253.

Marx, A. et al. (2012), *Private Standards And Global Governance, Economic, Legal and Political Perspectives*, Edward Elgar Publishing.

Mattli and Woods (eds) (2009), *The politics of Global Regulation*, Princeton University Press.

Memorandum of Understanding between Internet Platforms and Rights Owners, http://ec.europa.eu/internal_market/iprenforcement/docs/memorandum_04052011_en.pdf, 4 May 2011.

Memorandum of Understanding on Working Arrangements Between Asian Development Bank and United Nations Environment Programme, www.adb.org/sites/default/files/mou-unep.pdf.

Memorandum of Understanding against Copyright Infringement, www.copyrightinformation.org/wp-content/uploads/2013/02/Memorandum-of-Understanding.pdf.

Merges, R. P. (1996), "A Comparative Look at Intellectual Property Rights and the Software Industry", in D. C. Mowery (ed.), *The International Computer Software Industry: A Comparative Study of Industry Evolution and Structure*, Oxford University Press.

Morgenstern and Pizer (eds) (2007), *Reality Check: The Nature and Performance of Voluntary Environmental Programs in the United States, Europe, and Japan*, RFF Press, Washington, D.C.

MoU between the Global Reporting Initiative (GRI) and the UN Global Compact (GC), GRI and UN Global Compact Forge New Alliance www.unglobalcompact.org/news/50-06-24-2010.

OECD (2007), "OECD Competition Assessment Toolkit 2007 (revised in 2010)", www.oecd.org/daf/competition/competitionassessmenttoolkit.htm.

OECD (2010), *Transition to a Low-carbon Economy: Public Goals and Corporate Practices*, OECD Publishing, Paris, www.oecd.org/daf/inv/mne/transitiontoalow-carboneconomy.htm, doi: 10.1787/9789264090231-en.

OECD and ISO (2008), "Memorandum of Understanding between the Organization for Economic Cooperation and Development (OECD) and the International Organization for Standardization (ISO) in the Area of Social Responsibility", www.oecd.org/dataoecd/51/11/45330481.pdf, 5 May.

OECD-Global Reporting Initiative "Memorandum of Understanding between the Organization for Economic Co-operation and Development (OECD) and the Global Reporting Initiative (GRI)" www.oecd.org/daf/internationalinvestment/guidelinesformultinationalenterprises/46936939.pdf.

Official Journal of the European Union (2010), "Regulation laying down the obligations of operators who place timber and timber products on the market", No. 995/2010, 20 October.

Sabel, C. F. and W. H. Simon (2006), "Gaps and Hybrids", in G. de Burca and J. Scott (eds), *New Governance and Constitutionalism in Europe and the US*, Hart Publishing, Oxford.

Sabel, C. F. and J. Zeitlin (2010), *Experimentalist Governance in the European Union: Towards a New Architecture*, Oxford University Press.

Savona, E. U. et al. (2006), "A Study on Crime Profiling – Evaluation of Crime Risk Implications of the European Commission's Proposals Covering a Range of Policy Areas", Final Report, http://transcrime.cs.unitn.it/tc/fso/pubblicazioni/AP/Final_Report-A_study_on_Crime_Proofing.pdf.

Schepel, H. (2005), *The Constitution of Private Governance – Product Standards in the Regulation of Integrating Markets*, Hart Publishing.

Schmidt, R. and P. Verbruggen (2013), "The Role of Certification", in F. Cafaggi (ed.), *Enforcing Transnational Private Regulation – A Casebook*, Edward Elgar Publishing.

Scott, C., Cafaggi, F. and L. Senden (eds) (2011), *The Challenge of Transnational Private Regulation: Conceptual and Constitutional Debates*, Wiley-Blackwell.

SOCHI 2014 and UNEP (2009), "Memorandum of Understanding between the Organizing Committee of XXII Olympic Winter Games and XI Paralympic Winter Games of 2014 in the City of Sochi (Sochi2014) and the United Nations Environment Programme (UNEP)", http://unep.org/sport_env/sochi2014/Documents/Other/UNEP-SOCHI%20MOU.pdf, 5 June.

Sorsa, K. (2011), "Private-Regulation in Global Value Chain – a Trade Barrier or an Opportunity for Public-Private Co-operation?", http://regulation.upf.edu/dublin-10-papers/1H4.pdf.

Stewart, R. (2012), "Enforcement of Transnational Public Regulation", in F. Cafaggi (ed.), *The Enforcement of Transnational Regulation: Ensuring Compliance in a Global World*, Edward Elgar, Cheltenham and Massachusetts, p. 41.

Stiglitz, J. E., A. Sen, J.-P. Fitoussi (n.d), "Report by the Commission on the Measurement of Economic Performance and Social Progress", www.stiglitz-sen-fitoussi.fr/documents/rapport_anglais.pdf.

Sustainability Impact Assessment (SIA) of the EU-ACP Economic Partnership Agreements (2006). Phase Three: Horticulture in Eastern and Southern Africa (ESA), Final Report, 11 September, PricewaterhouseCoopers, http://trade.ec.europa.eu/doclib/docs/2006/september/tradoc_130129.pdf.

UNEP and EPA (2011), "Memorandum of Understanding between the United Nations Environment Programme (UNEP) and the Environmental Protection Agency of the United States of America (EPA)", www.epa.gov/international/io/epaunepmou.pdf, 21 February.

UNEP and FELEBAN (2011), (Memorandum of Understanding between The United Nations Environment Programme (UNEP) and the Federación Latinoamericana de Bancos (FELABAN), www.felaban.com/convenios/unep/memorando_unep.pdf.

UNGC and EFQM (2003), "Memorandum of Understanding between the United Nations Global Compact (GC) and the European Foundation for Quality Management (EFQM)", www.unglobalcompact.org/docs/news_events/9.1_news_archives/2003_10_30/mou_efqm.pdf, 30 June.

United Nations Global Compact (2010), "GRI and UN Global Compact Forge New Alliance", www.unglobalcompact.org/news/50-06-24-2010, 24 June.

United Nations Global Compact Office and ISO (2006), "Memorandum of Understanding between the Global Compact Office and the International Organization for Standardization (ISO)", www.scribd.com/doc/16810157/20061109-Memorandum-of-Understanding-between-ISO-and-Global-Compact.

Van Waarden, F. (2010), "Governing Global Commons: Public-Private-Protection of Fish and Forests", *Jerusalem Papers in Regulatory Governance*, paper No. 17, June.

Waddell, S. (2011), "The Global Compact: An organizational innovation to realize UN Principles", www.unglobalcompact.org/docs/news_events/9.1_news_archives/2011_11_16/UNGC_Organizational_Innovation_Note.pdf.

WalMart (n.d.), "WalMart Sustainability Index", www.walmartstores.com/Sustainability/9292.aspx.

Wood, S. and L. Johannson (2008), "Six Principles for Integrating Non-Governmental Environmental Standards into Smart Regulation", Vol. 46, *Osgoode Hall Law Journal*, pp. 345-396.

Zimmermann, J., J. Werner and P. Volmer (2008), *Global Governance in Accounting: Rebalancing Public Power and Private Commitment*, Palgrave Macmillan.

Chapter 2

Transboundary water management

by

Julia Black and Céline Kauffmann*

There are 261 transboundary river basins in the world, representing 45% of the earth's land area. Nineteen basins cross five countries or more, including the Mekong, the Nile, the Niger and the Rhine. The Danube, for example, flows adjacent to, or through 18 countries. In Europe alone, 20 countries rely on neighbouring countries for more than 10% of their water resources and five European countries draw 75% of their water resources from upstream countries. This case study focuses on modes of international regulatory co-ordination in water governance, specifically in managing river basins that cross national boundaries for non-navigational purposes.

* Julia Black is Professor of Law at the London School of Economics and Political Science. Celine Kauffmann is senior economist in the OECD Regulatory Policy Division. Comments and references were provided by Aziza Akhmouch of the OECD. Flavia Donadelli provided research assistance.

Introduction

This case study focuses on modes of international regulatory co-ordination (IRC) in water governance, specifically in managing river basins that cross national boundaries for non-navigational purposes.[1] This is a significant matter: there are 261 transboundary river basins in the world, constituting 45% of the earth's land area (UNDP, 2004). Globally, 19 basins cross 5 countries or more, including the Mekong, the Nile, the Niger and the Rhine. The Danube, for example, flows adjacent to or through 18 countries. In Europe alone, 20 countries depend for more that 10% of their water resources on neighbouring countries and 5 European countries draw 75% of their resources from upstream countries (UNESCO, 2003).

Rivers that flow across national boundaries create significant interdependencies between the riparian countries through which they flow. In particular, those down-river are vulnerable to the activities of those up-river in a variety of significant ways, from over-extraction of water or the building of dams (so depriving countries down-river of water), or from pollution and water-borne diseases (so depriving those down-river of clean, safe water). Conversely, activities down-river can be a contributory cause of flooding up-river.[2]

These interdependencies can intensify the competition for water between riparian countries, but can also promote the search for co-ordination. Since 1948, there have been only 37 incidents of acute conflict over water, while during the same period, approximately 295 international water agreements were negotiated and signed (UN-Water, 2008).[3] However, these agreements are often incomplete or not respected, as discussed below.

Given the characteristics of the sector, as well as the number of actors involved, water governance is already a complex and challenging undertaking at domestic level (OECD, 2009; 2011). Those complexities and challenges multiply and intensify when attempted on a cross-border basis, independently of any wider political conflicts between the countries concerned (UNDP, 2004; Timmerman and Bernardini, 2007). Challenges arise from the differences between priorities and needs of different countries; their national water management regimes (the principles, rules and procedures that steer water management); and differences in governance capacity. Transboundary water management requires co-ordination over different political, legal and institutional settings as well as over different information management approaches and financial arrangements (Timmerman and Bernardini, 2007). It is often hampered by weak social and institutional capacity, and poor legal and policy frameworks at the national level, further amplified by contrasting levels of knowledge, capacities and institutional frameworks between riparian states, in the context of potential inconsistency and conflict of policies. Finally, the longer term benefits of co-ordination can be overridden by the play of short term interests.

Transboundary co-ordination at the river-basin level is frequently supported by a broader infrastructure of institutions operating at the global and regional level. The first section of the case study therefore provides an overview of the principal organisations involved in co-ordinating transboundary water governance at the global level and the key legal instruments in place. It gives some examples of governance structures for some of the major river basins in the world, and examples of co-ordination mechanisms used. The second section focuses on benefits, challenges and costs of regulatory co-operation in managing transboundary water. The third section explores one example in more depth, the International Commission for the Protection of the Rhine.

An overview of mechanisms and actors

In practice co-operation is based on international customary law and on a wide range of formal and informal agreements.

Legal principles and agreements

The legal principles governing transboundary water management are set out in a series of UN conventions, bi-lateral and multi-lateral agreements between states, in the case law of the International Court of Justice and in international customary law. The two key Conventions setting out the legal principles on transboundary water management are generally regarded to be the 1992 Helsinki Convention (Box 2.1) and the 1997 UN Convention (Box 2.2) though the latter is still non-binding. In addition, a study by the FAO finds over 3600 water-related treaties, dating back to 805AD. Since the Second World War, over 250 legally binding agreements have been reached with respect to individual river basins (Wolf, 1997). Most are bilateral,[4] but important multilateral agreements exist for the management of some significant watercourses, including the Nile, Niger, Ganges, Mekong, Colorado, Danube, and Rhine rivers. Details of the key legal instruments operating at the global and regional level are set out in Annex 2.A1; examples of binding agreements with respect to individual river basins are set out in Annex 2.A2.

Co-ordination between countries over transboundary or boundary-defining watercourses is also based on non-binding political agreements, and historical and customary use. According to a recent UN report, 158 of the world's 263 international river basins, plus transboundary aquifer systems, still lack any type of co-operative management framework or agreement (UN-Water, 2008).

Box 2.1. Helsinki Convention 1992

The Convention sets out a minimum framework for agreements between riparian states on the management of transboundary watercourses based on co-operation, equality and reciprocity, good faith and good-neighbourliness. The EU and 36 countries have ratified: the EU member states (excluding Ireland, Cyprus* and Malta) plus Albania, Azerbaijan, Belarus, Bosnia and Herzogovena, Croatia, Kazakhastan, Liechtenstein, Norway, Moldova, Russian Federation, Serbia, Switzerland, Ukraine and Uzbekistan. Activities under the Convention are supported by the UNECE Secretariat.

The Convention (Article 1) provides that all parties "shall take all appropriate measures to prevent, control and reduce any transboundary impact", at source wherever possible. In particular, the parties shall take all appropriate measures to ensure that transboundary waters are used in a reasonable and equitable way; to prevent, control and reduce pollution; to ensure that transboundary waters are used with the aim of ecologically sound and rational water management; and to ensure conservation and, where necessary, restoration of ecosystems.

In so doing, the Convention provides that the parties shall follow 3 principles familiar to environmental regulation in the EU:

- the precautionary principle, by virtue of which action to avoid the potential transboundary impact of the release of hazardous substances shall not be postponed on the ground that scientific research has not fully proved a causal link between those substances, on the one hand, and the potential transboundary impact, on the other hand;
- the polluter-pays principle, by virtue of which costs of pollution prevention, control and reduction measures shall be borne by the polluter; and
- the inter-generational principle, which provides that water resources shall be managed so that the needs of the present generation are met without compromising the ability of future generations to meet their own needs.

Box 2.1. **Helsinki Convention 1992** (*cont.*)

Emphasis is placed on the need for information exchange, the development of common systems of data collection, analysis and presentation, joint research into and development of effective techniques for the prevention, control and reduction of transboundary impact, and the development of compatible regulatory measures to control pollution. In addition, parties are to provide warning and alarm of critical situations and provide mutual assistance in a crisis upon request. The Convention is a framework document which also provides that institutional structures in the form of joint bodies should be put in place to implement bilateral or multi-lateral agreements between riparian states. These currently include joint bodies for the management of the Danube, Elbe, Meuse, Moselle and Saar, Odura, Rhine, Aral, Chu-Talas, Lake Constance, Lake Geneva and the Saar. The Convention also has its own institutional structure. The supreme decision making body is the meeting of the parties, held every 3 years. The day to day work is conducted by the Bureau (3 members elected by the parties), working groups and a small secretariat. The work is supported by the UNECE, which together with other partners conducts and facilitates several projects including assessments of transboundary waters, projects on co-operation notably in Eastern Europe, the Caucasus and Central Asia, and facilitates national policy dialogues. The UNECE produces monitoring and progress reports on particular aspects of co-operation but there is no overall body to which the separate bilateral or multilateral bodies have to report or to which they are held accountable.

Disputes between parties are resolved either by the ICJ or through arbitration under the convention. These provisions are incorporated into the river-specific co-operation agreements, and it is up to the parties to the dispute which route to take in any one case.

*. Note by Turkey:

The information in this document with reference to "Cyprus" relates to the southern part of the Island. There is no single authority representing both Turkish and Greek Cypriot people on the Island. Turkey recognises the Turkish Republic of Northern Cyprus (TRNC). Until a lasting and equitable solution is found within the context of the United Nations, Turkey shall preserve its position concerning the "Cyprus issue".

Note by all the European Union Member States of the OECD and the European Commission:

The Republic of Cyprus is recognised by all members of the United Nations with the exception of Turkey. The information in this document relates to the area under the effective control of the Government of the Republic of Cyprus.

Source: Based on the Helsinki Convention on the Protection and Use of Transboundary Watercourses and International Lakes (1992), www.unece.org/fileadmin/DAM/env/water/pdf/watercon.pdf.

Even where legal agreements are in place, however, these may be ineffective for a number of reasons. Three legal conventions and a commission did not prevent widespread pollution of the Rhine, for example (Schwabach, 1989). They may be partial in that they only relate to certain aspects of transboundary water management (for example, hydro power and water supply but not environmental protection), or because important riparian countries are not party to the agreement (e.g. Turkey is not party to the agreement with respect to the Orontes river which flows through Turkey, Syria and the Lebanon);[5] or because the legal agreement is widely ignored and there is no effective enforcement mechanism (Wolf, et al. 1999; ODI, 2001; UN-Water, 2008). Wolf and Hamer (2000) found that of most treaties analysed, more than half include no monitoring provisions whatsoever, two-thirds do not delineate specific allocations and four-fifths have no enforcement mechanism.

Box 2.2. 1997 UN Convention on the Law of non-Navigational Uses of International Watercourses

The UN Convention sets out the principles of utilisation of joint watercourses as established in international customary law. The Convention has not yet been ratified by the required 36 countries to come into force, although negotiations are on-going.[6] It does not completely overlap with the Helsinki Convention, in particular there is no mention of the precautionary principle, polluter-pays principle or the inter-generational principle. It is also wider in scope, in that it focuses not only on pollution but also on the protection and preserve ecosystems in general, including the prevention of the introduction of alien or new species which could have a significantly harmful effect on the ecosystem. However, it contains a similar set of obligations on co-operation and on the day to day management of transboundary watercourses, including information exchange, notification of planned activities and crisis management.

The key elements of the Convention are the following:

- The principle of equitable and reasonable utilisation and participation, whereby participation includes both the right to utilise the watercourse and the duty to co-operate in its protection and development.
- The no-harm principle: the obligation not to cause significant harm, and where harm does occur an obligation to eliminate or mitigate such harm and, where appropriate, to discuss the question of compensation.
- The principle of equality of priority of use/no use of an international watercourse enjoys inherent priority over other uses in the absence of custom or other agreement.

Box 2.2. **1997 UN Convention on the Law of non-Navigational Uses of International Watercourses** (*cont.*)

- The duty to co-operate in the management of the watercourse on the basis of sovereign equality, territorial integrity, mutual benefit and good faith in order to attain optimal utilisation and adequate protection of an international watercourse, including the establishment of joint commissions.
- The requirement for notification and advance warning: to give information to other riparian states concerning planned measures and to co-operate in their implementation where appropriate, and to notify other riparian states concerning planned measures with possible adverse effects, and a process to be followed for reply and negotiation.
- The requirement to co-operate in the management of emergency situations and to develop joint contingency plans.

In addition, the Convention imposes a duty to exchange readily available data and information on the watercourse and related forecasts on a regular basis, with provision for the defraying of reasonable costs in its acquisition at the request of another party. It also requires parties to protect and preserve ecosystems; prevent, reduce and control pollution; to take all necessary measures to prevent the introduction of alien or new species which could have a significantly harmful effect, and to protect and preserve the marine environment. It imposes an obligation to, individually and, where appropriate, jointly, take all appropriate measures to prevent or mitigate conditions related to an international watercourse that may be harmful to other watercourse States, whether resulting from natural causes or human conduct.

Source: Based on United Nations Convention on the Law of Non- navigable Uses of International Watercourses http://untreaty.un.org/ilc/texts/instruments/english/conventions/8_3_1997.pdf.

Furthermore, as research and technologies of water governance develop, agreements can become outdated. Only the most recent ones, such as the EU Water Framework Directive, adopt an integrated approach to river basin management or include concern for the ecosystem in their provisions, for example. Wolf (1999) established that although it is regarded as a statement of existing international customary law, many of the provisions of the UN Convention are missing from existing (often earlier) transboundary agreements, and even where they are present, the principles themselves conflict (Wolf, 1997). In particular, the "no harm" principle operates in tension with the principle of "reasonable and equitable use". However, the draft provisions of the 1997 Convention have formed the basis for the

Convention on the Danube in 1994 and the reconstitution of the Mekong River Commission in 1995. Nonetheless, most agreements deal with highly localised issues, which vary with the hydrological, political and cultural context of the different river systems.

Institutional arrangements in support of IRC

At the international level, there is a legal hierarchy of norms created through the Conventions and associated Protocols; to this extent the authors of the Protocols are "leaders" setting standards that others implement. In practice, co-operation is through bilateral or peer group arrangements, or through managed networks.

The UN has a significant role in promoting and facilitating effective water governance internationally through the activity of some 28 UN agencies. The UN organises itself through a series of managed networks of UN related agencies and programmes, co-ordinated through UN-Water (see Annex 2.A3), and operates with slightly different sets of institutional partners with respect to different programmes, both other international governmental organisations and international non-governmental organisations (INGOs). At the regional level, the regional UN bodies also partner with a wide range of industry and civil society organisations on different initiatives (see Annex 2.A4 for a sample of non-governmental organisations involved in trans-boundary water management). Frustration with the fragmented landscape of water management at international level is leading some to consider the creation of a single UN body exclusively dedicated to water, although this option is raising a number of oppositions.[7]

Agreements with respect to individual river basins may be accompanied by the establishment of an institutional structure to implement the agreement, such as the International Committee for the Protection of the Rhine; the International Committee for the Protection of the Danube; the Mekong River Commission; the Nile Basin Initiative (see Box 2.3); or the Volta River Basin Organisation. In practice, where institutional arrangements exist, they vary in a number of respects as they are closely linked to surrounding political environments, and are thus sensitive to changes in those environments (ODI 2001; ECOWAS – SWAC/OECD, 2006).

Often, there is no institutional structure to manage co-ordination. For example in the central Arab region, in some instances co-operation is formalised by inter-state agreements, some less formally set up through technical committees, experts meetings, or joint projects. Numerous shared water basins are still managed in a unilateral manner by the concerned states, without any co-operative effort. Matters are rendered more complicated where there is longstanding conflict between riparian states on wider issues. There is no co-operative agreement on the management of the River Jordan between the riparian countries, for example, with the exception of the 1994 Treaty of Peace between Israel and Jordan on water allocation (AFED, 2010). Even where co-operative modalities exist in exchanging data and developing models and information systems, actual joint management of the shared water systems has not yet taken root or is held up by conflict (AFED, 2010).

Box 2.3. **The Nile Basin Initiative**

The Nile Basin Initiative launched in 1999, for example, brought together the 10 riparian countries of the Nile (Egypt, the Sudan, Burundi, Democratic Republic of Congo, Eritrea (observer), Ethiopia, Kenya, Rwanda, Tanzania, and Uganda). The objectives of the initiative are to develop the river in a co-operative manner, share substantial socioeconomic benefits, and promote regional peace and security, and it has initiated seven projects in the area pursuing different aspects of transboundary water management. The NBI is comprised of a council of ministers of water affairs of the Nile Basin (Nile-COM), a technical Advisory Committee (Nile-TAC), and a secretariat (Nile-SEC). However difficulties still remain with respect to 3 important areas: protection of the historical share of the Nile water for the downstream countries (Egypt and the Sudan claim preferential rights based on agreements in 1927 and 1957), governance, and the proposal for unanimous agreement for decision making especially in approving investment operations in the basin. Under the NBI initiative, a Co-operative Framework Agreement (CFA) was prepared, based on the 1997 UN Convention, and opened for signature in 2010. This would create the Nile Basin Commission. Seven states have ratified, so the agreement has come into operation, but Egypt, the Sudan and the Democratic Republic of Congo are still opposed.

Main co-operation mechanisms used

Instruments of co-ordination

In addition to the agreements and institutional agreements discussed above, the main co-ordination mechanisms used at the global level are the following:

- Membership in international organisations promoting regulatory co-operation (e.g. UNDP, UNESCO, WHO, or the Global Water Partnership);
- Facilitating formal (umbrella type) regulatory co-operation partnerships between countries, principally through the regional UN bodies (e.g. UNECSWA, UNECE);

- Facilitating the creation of river- basin co-operative agreements, including the creation of a specific organisation charged with monitoring and implementing the agreement;

- Provision of financial support for co-ordination, e.g. through the Global Environmental Fund and the World Bank or through individual countries support;

- EU "conditionality" policy in the allocation of regional structural funds under the Water Framework directive, which requires riparian states to co-ordinate in the management of transboundary water, for example with respect to the Danube or Sava.

Functions being co-ordinated/components covered in agreements

In an analysis of 49 treaties, Wolf (1997) finds that there is little common ground between them; instead they tend to reflect the often unique setting and needs of each basin and surrounding political context. In particular, they vary as to the parties (whether they are bi-lateral or multi-lateral, and include all riparian states or only some); as to subject matter (ranging from data collection to allocation, planning, construction, ecological management); territorial extent (whether they cover the whole basin on only part of it) and intensity of co-operation (from duties to inform to development and implementation of joint programmes) (Kliot et al., 2001).

The most complete agreements (e.g. the ICPR) follow the 1992 and 1997 Conventions in covering all of the following:

- *Ex ante* exchange of information and development of common systems of data collection, measurement and assessment;
- Agenda setting/setting goals;
- Formulating rules/norms/standards;
- Monitoring, data collection;
- Supervision;
- Enforcement;
- Crisis management;
- "Clean up"/responding to disasters which have cross-border dimensions.

Other issues covered include:

- water allocation;
- construction of dams;
- joint research projects;
- capacity building mechanisms.

A frequently observed barrier to co-ordination is the lack of a common dataset on which riparian countries can agree (ODI, 2001 and Frijters and Leentvaar, 2003). The absence of a common baseline agreement on critical facts such as river flow or pollution levels can be a significant impediment to transnational co-operation. Part of the problem is the lack of a common set of terminology and common systems of data assessment and measurement. To this end there have been a number of global attempts to create common terminology and methods of water accounting. According to UN-Water (2008), these include:

- UNESCO and the World Meteorology Organisation's International Glossary of Hydrology, which translates an international standard vocabulary into regionally used languages;
- United Nations Statistics Division and the Division for Sustainable Development of UN-DESA's System of Environmental-Economic Accounting for Water;
- UNESCWA Arabic glossary on transboundary water.

Overall assessment and next steps

Known benefits

UN-Water (2008) argues that co-ordination can, directly or indirectly, contribute to international trade, economic development, food security, political security, poverty alleviation and regional integration in a number of ways. More specifically, a range of studies has found that co-ordination in water governance can have the following benefits (UN Water, 2008, AFED Report, 2010):

- progress in managing risks across borders (in relation to the environment, human health, sustainable development and human security, including loss of life in the event of floods);
- better ecological management and environmental sustainability, providing benefits to river, aquifer, lake, wetland and related ecosystems as well as adjacent estuaries, coastal areas and seas;

- increased food and energy production;
- poverty reduction and socio-economic development;
- help to control migration;
- greater transparency and work-sharing across governments and public authorities, which can help to reduce wider tensions between riparian countries;
- improved economic integration between co-ordinating states (though in some cases this may need to be in place to make the co-ordination work.

Challenges

Water governance raises a number of complex and challenging issues even at the national level, which intensify in the context of transboundary co-ordination. The difficulties are in part a function of the size and complexity of the river basin in question. For example, a river such as the Niger is more than 4 000 km long, and has a basin covering 1.5 million km^2 which is shared between nine countries (ECOWAS – SWAC/OECD, 2006). The nature of the water basin itself imposes considerable demands on the co-ordinating capacities of the riparian countries. In other cases, where there are only two riparian countries, and where the shared water resource is not critical to either of them, co-ordination is likely to be easier.

However, a shared watercourse is often a scarce resource over which riparian countries compete. In these contexts, whilst all may sign an agreement confirming the right of each to an "equitable allocation" of water, just what this means can be highly contested. This is particularly the case where there are considerable differences in economic development between them (e.g. in the case of the Mekong) or in situations where one country has historically utilised all the flow and claims prior rights (e.g. the Jordan and the Nile) (ODI, 2001). It is, however, also the case where the riparian countries are well developed, as illustrated by the disputes between the US and Canada.[8] Moreover, not all riparian countries may be party to the agreement: China and Burma are not members of the Mekong River Commission, for example, although they are "dialogue partners".

Some of the most critical challenges, based on OECD (2011), UNDP (2004), ODI (2001), Timmerman and Bernardini (2007), AFED (2010) and Jacobs (1995) include:

- Weak governance capacity at the national level:
  - lack of co-ordination between parts of the national government with responsibilities related to water (e.g. between ministries, departments or agencies for agriculture, environment, trade, health and sanitation, energy, regional or sectoral development, and planning);
  - lack of co-ordination between donor agencies (in developing countries);
  - lack of co-ordination between private and public bodies with responsibilities for water provision and water governance;
  - lack of personnel with the requisite skills and knowledge;
  - lack of adequate data collection, monitoring and enforcement systems;
  - lack of adequate financing.
- Differences between riparian countries in terms of:
  - socio-economic development;
  - governance capacities;
  - technical infrastructure;
  - legal frameworks;
  - political and cultural orientations;
  - data collection, measurement and assessment, preventing the establishment of baseline data on which to base agreements.
- Unequal distribution of costs and benefits within and across countries:
  - competition for a scarce resource;
  - conflicts of interest and priorities;
  - negative externalities of the actions of one riparian country on another, for example construction of dams or hydro-electric power stations, pollution.
- Broader political tensions, often historically entrenched, including armed conflict (UN-Water, 2008).

Requirements of successful co-ordination

Developing a framework between riparian countries for the joint management of their shared water resources is an intense, primarily political process of building relationships and trust through dialogue (UNDP, 2004). The distinct differences between river basins in terms of their physiography, climate, hydrology, natural resources and the dynamics of their surrounding eco-systems, which often vary across the river basin, and between riparian states in terms of their political, social, cultural, historical and economic contexts, mean that each river basin is unique. It is thus neither possible nor desirable to draw up a detailed blueprint for co-ordination which can be applied to every transboundary river.

Nevertheless, certain principles of successful co-ordination recur in many of the existing studies on transboundary water co-ordination – Jacobs, 2002; S. El-Din Amer, Y. Arsano, A. El-Battahani, O. El-Tom Hamad, M. Abd El-Moenim Hefny5 and I. Tamrat, 2005; Wolf, 1997; Komakech, Jaspers, van der Zaag, 2007; and Mostert, 2003. They are:

- Political context
  - Political commitment by all riparian states;
  - Political stability at the national level and commitment to effective water governance;
  - Mutual recognition of claims and shared commitment to co-ordination in water governance between all riparian states (ODI, 2001);
  - Sufficiently common goals and agenda between riparian states;
  - Disasters provide important triggers for co-ordination, or enhance existing mechanisms.[9]
- Governance capacity at the national level, including integrated approach to water management
- Strong institutional structure at the transboundary and regional level including:
  - creation of common systems of data collection, measurement, assessment and a common terminology – a number of studies have found that the exercise of building common data sets and engaging in joint monitoring can improve co-operation and co-ordination, even when there are wider political tensions (AFED, 2010; Mamou *et al*, 2006; Huisman *et al*, 2000; Savenijeand Van Der Zaag, 2000);

  - strong legal framework for agreements between riparian states (Kliot, Shmueli, and Shamir, 2001) which clearly set out duties and responsibilities of the parties and incorporate systems for enforcement and dispute resolution;
  - creation of a multi-member river-basin institution involving riparian states to implement the agreement with a clear mandate, adequate funding, backing from all member states, and strong powers for joint monitoring, information exchange, obligations of mutual notification and assistance, enforcement and dispute resolution mechanisms;
  - public participation and co-operative working with other bodies at the governmental and non-governmental level and major stakeholders including funding bodies, research organisations, NGOs, local communities and civil society groups, and individual water users and/or influential individuals at the local level (Newton undated);
  - holistic and integrated approach to water governance, recognising social, cultural, ecological, environmental, economic and technical/engineering issues;
  - co-operation with other joint bodies in the same area with different scope (e.g. navigation, aquifers) or with same scope in different areas (e.g. protection of inland waters or marine environment);
  - long term and contingency planning, including creation of systems of resilience (e.g. flood defences; crisis management and clean up mechanisms in cases of extreme and accidental pollution);
  - periodic reviews and assessments followed by adaptations and modifications in goals or strategies to meet changing circumstances or opinions (Huisman et al., 2000);
  - Strong links between national, regional and river-basin levels of government, to build capacity and ensure appropriate co-ordination, implementation and adaptation.[10]

- Appropriate human capacity (UN-Water, 2008) and ownership/awareness
  - multidisciplinary teams to raise understanding of the complexities of managing shared water resources and derive the benefits made possible through co-operation;

- personnel with negotiation, diplomacy and conflict resolution skills as well as technical and scientific skills;
- development of relations of trust between delegates of riparian states;
- Creating awareness at the national level across all levels of government and with individual water users of the long term benefits of co-operation.

- Creation of a system for sharing costs and benefits, including payments where necessary (UN-Water, 2008; Dieperink, 2011).
  - Developing a consensus over basic entitlements;
  - Creating a development and management plan which pays attention to the differential distribution of costs resulting from the use of water resources over the entire river basin and which maximises overall benefits, though difficult trade-offs are likely;
  - Payments for benefits (or compensation for costs)[11] Special approaches may be needed with respect to benefits and costs that are not easily quantifiable or commensurable, such as for flood mitigation, regulating run-off and water supply.[12]
- Adequate financing: the effective development and maintenance of transboundary co-operation requires considerable financing, both to cover the operating costs of the institutional structure but more substantially to finance the measures which need to be taken. As an example, the cost of the ICPR Action Plan on Floods is estimated at EUR 18 billion (see www.iksr.org/index.php?id=123&L=3). In the case of developing countries this is often beyond the resources of the individual riparian countries so external funding is required, such as that provided by the UN Global Environment Fund, or by bilateral or multilateral donors.[13]
- Third party facilitation and support, particularly where there is tension over the use of shared water resources. Transboundary water management has been described as an "international public good", as such requiring international support (ODI, 2001; Wolf and Newton, n.d.; Jacobs, 1995).
  - Third parties can facilitate the creation of dialogue and mutual trust between countries, e.g. by providing neutral fora for information exchange;

  - Third party negotiation and brokerage of agreements should be conducted by organisations which are sufficiently strong and respected, e.g. the UN agencies, or by regional bodies, or by effective and well-respected river basin institutions;
  - Building broad partnerships between riparian countries and international donors to ensure co-ordinated programme implementation (UNDP, 2004).

Costs

There are at least two main sets of costs of co-ordination:

- Direct administrative costs
  - of IRC infrastructure, e.g. the river-basin institution;
  - of bi-lateral or multi-lateral liaison in the absence of a river basin institution.
- Costs arising from the substance of the co-ordination agreement itself: e.g. requirements to compensate riparian states for externalities imposed on them, or costs arising from decreased water allocation due to rebalancing of rights between riparian countries.

A wide variety of funding mechanisms exists, but there is continued concern at the levels of financial support available, in the co-ordination of funding, and in managing financial risks to donors and recipients (Joyce and Granit, 2010).

Next steps

The next stage in transboundary water co-ordination relates to co-ordination on groundwater aquifers. This is an issue of increasing international concern, which translated in 2008 in the development of a set of draft articles draft articles for the management of transboundary aquifers by the International Law Commission.

The mechanisms of co-ordination continue to be dialogue, information exchange, the conclusion of agreements, and are increasingly focused on efforts to build strong multilateral and inclusive institutions to manage individual river basins, for example the Mekong, the Nile and the Danube. Significant challenges remain, however, not least the difficulties of ensuring adequate political, institutional and financial support for co-ordination arrangements, particularly but not uniquely in the case of developing countries.

A case in point: The International Commission for the Protection of the Rhine

Main characteristics

The co-operation is taking place at the level of the Rhine river basin, including its alluvial areas and the waters in the watershed.

Actors involved

The International Commission for the Protection of the Rhine (ICPR) was formed in 1950 on a diplomatic basis between Switzerland, the Netherlands, France, Germany and Luxemburg. It was given a legal basis by the Berne Convention in 1963. The EEC joined as a member in 1976. The Berne Convention was revised and updated by the Rhine Convention in 1999 to take into account the UNECE-Helsinki Convention of 1992, the 1997 UN Convention and the EU Water Framework Directive. Since 2000, the ICPR Member States also co-operate on an equal basis with Austria, Liechtenstein, Italy and the Belgian region Wallonia which all have shares in the Rhine catchment, as required under the EU Water Framework Directive (ICPR, 2010).

The ICPR combines political representatives and technical experts (ICPR, 2004):

- Ministers meet every 2-3 years to set_common goals and agenda for the Commission, and the Presidency rotates every 3 years;
- Senior officials meet in plenary session on an annual basis to determine programmes, finances and procedures;
- A co-ordination group meets four times a year and is responsible for planning and co-ordination of the work of the ICPR;
- There are a number of permanent working groups, and individual project and expert groups on separate issues;
- The ICPR is supported by a small secretariat based in Koblenz, Germany;
- River commissions from the region have observer status in plenary sessions;
- The ICPR recognises NGOs as eligible to observe in plenary meetings and to participate in working groups as observers or experts.

The International Commission for the Hydrology of the Rhine Basin was founded in 1970 to promote co-operation between riparian states. It is a permanent, autonomous, international commission and has the status of a foundation registered in the Netherlands. The members of the Commission are the scientific and operational hydrological institutions of the Rhine basin. Since 1975, it has worked within the framework of the International Hydrological Programme of UNESCO and the Hydrology and Water Resources Programme of the World Meteorological Organisation (WMO). However, it has not been granted observer status to meetings of the ICPR. In principle, its functions could be performed by the IRC, but the ICHRB is an established body with a longer history of successful co-operation.

Intended objectives

The 1999 Convention sets out six aims (Article 3):

- sustainable development of the Rhine ecosystem;[14]
- the production of drinking water from the waters of the Rhine;
- improvement of sediment quality in order that dredged material may be deposited or spread without adversely affecting the environment;
- general flood prevention and protection, taking account of ecological requirements;
- help restore the North Sea.

To these ends, the Convention requires members to be guided by the following principles:

- precautionary principle;
- principle of preventive action;
- principle of rectification, as a priority at source;
- polluter-pays principle;
- principle of not increasing damage;
- principle of compensation in the event of major technical measures;
- principle of sustainable development;
- application and development of the state of the art and best environmental practice;
- principle of not transferring environmental pollution from one environment to another.

Form that the co-operation is taking

- *Formality* – legally binding (in addition to the protection of the Rhine, the Commission follows the EU Water Framework Directive and EU Floods Management Directive);
- *Scope* – Comprehensive and holistic in environmental terms;
- *Mode of co-ordination:*
  - Consensus within the Commission – all decisions within the Commission are to be taken on the basis of unanimity.
  - Authority delegated to Commission by signatories, who then have to abide by Commission decisions, implement them in accordance with their national laws, and report to the Commission on implementation (1999 Convention, Article 11).
- *Instruments of co-operation:*
  - Membership in international organisation (ICPR) established by an international agreement.
- Functions being co-ordinated/components covered in agreements
  - *Preparation of measuring programmes and studies* – use of results if necessary (data collection/research);
  - Make proposals for individual measures and programmes of measures (setting goals);
  - Co-ordinate the Contracting States' warnings and alert plans for the Rhine (supervision) – Alarm Model for the Rhine;
  - *Evaluate* the effectiveness of the actions decided upon (monitoring);
  - Carry out any other tasks entrusted to it by the Contracting Parties. According to Huisman et al. (2000) this last task proved to be very useful because the Rhine states could charge the ICPR with the rehabilitation of the Rhine ecosystem in 1987 and the food problems in 1995.

Short history of the development of the ICPR

The Rhine has a long history of conflicts and international agreements, from the Peace Conference in Vienna 1815, where the Rhine riparian states agreed to remove the river's tolls and improve navigability and created the Central Commission for Navigation on the Rhine, the oldest still active river commission in the world, until the recent 1999 Convention on the Protection of the Rhine (Bernauer and Moser, 1996; Frijters and Leentvaar, 2003; Huisman, Jong and Wieriks, 2000; and Dieperink, 2011). A short chronology is set out in Annex 2.A5.

A number of events have acted as triggers for co-operation at different stages over time, successively depletion of salmon stocks; navigation; pollution; downstream salination; fire; and floods.

In 1850, overfishing of salmon of the Rhine became a real threat to the sustainability of the species. As a result a salmon treaty was conducted in 1869, which was rejected by the Dutch government. A further treaty was agreed by all states in 1885, but over time navigation and hydro-power proved to be stronger concerns, and salmon gradually became extinct. In 1919, France imposed a provision in the Treaty of Versailles to allow the construction of a lateral canal along the Rhine's banks. Soon after the Second World War, Germany and France agreed a modification to the Treaty to allow the canalisation of the river and the construction of lateral channels, a solution which served both countries' interests.

Industrialisation along the banks of the Rhine was causing increasing pollution from the 1930s onwards. Wastewater discharges by industries, agriculture, traffic, and households resulted in significant amounts of heavy metals, pesticides, hydrocarbons, and organic chlorine compounds being discharged into the river, causing severe eco-toxicological problems. They also started to cause a salinisation problem at the Netherlands. After many years, the Dutch government succeeded in drawing attention of the other riparian states to the problem of pollution and, in 1950, the Federal Republic of Germany, France, Luxembourg, the Netherlands and Switzerland created the International Commission for the Protection of the Rhine against Pollution (ICPR). Initial work focused on establishing a common basis for assessing the river quality.

In 1963, the Commission received its legal foundation with the Convention of Berne (Huisman et al., 2000). The 1963 Convention did not set out any rules or procedures for decision making, however. Contentions continued to exist, particularly with respect to pollution. Due to the deposit of oxygen-consuming waste into the Rhine all aquatic life disappeared from the downstream sections of the river. This generated significant public concern, and in 1972 the Rhine states charged the ICPR to elaborate a convention to reduce chemical pollution. The Convention on Chemical Pollution, signed by the ICPR contracting parties in 1976, was an outline Convention that, among other provisions, provided for threshold values for the discharge of individual toxic substances into the environment, but exact emission standards were left to be agreed. The Convention identified 83 chemicals for which standards had to be agreed as a matter of priority; due to ongoing disputes standards for the first twelve of these were only agreed in 1986.

Further, activities upstream, in particular waste discharges from French potassium mines, were causing significant salination of the water downstream, in the Netherlands. In 1976 the Convention for the Protection of the Rhine against Chloride Pollution was signed. Under this agreement, the salt from mining activities had to be stored at the French potassium mines. Later, in 1991, a protocol to this "Salt Treaty" was signed which stipulated additional storage measures whenever the chloride level at the Dutch/German border reached a certain level. However, negotiations stalled on the issue of responsibility for payments for the storage and clean-up measures, and were eventually settled subsequent to arbitration in 1995 (Dieperink, 2011).

A further important trigger of the transboundary co-operation was the Sandoz disaster in 1986. A warehouse at the Sandoz chemical industries, near Basle, caught fire. The fire was extinguished with water from the Rhine that mixed with the chemicals and went back to the river. Tons of fishes and other animals died. The incident generated a great deal of public concern and the result was the Rhine Action Programme (RAP), agreed in 1987. The main objectives of the RAP were the return of many higher species (such as salmon) to the river by 2000; the future use of Rhine water for public water supply using simple purification techniques and the reduction of pollution of sediments.

In 1995, significant flooding along sections of the Rhine prompted further co-operation on floods and on improving the alarm system that was already in place. An "Action Plan on Floods" was passed by the Rhine ministers in 1998.

Most recently, in 2001, the "Rhine 2020 Programme" or the "Programme on the Sustainable Development of the Rhine" was adopted following the conclusion of the Rhine Action Plan (1987-2000). Among the core objectives of the "Rhine 2020" are:

- the implementation of Rhine habitat patch connectivity;
- Salmon 2020 – a programme which aims to ensure that there is a stable wild salmon population in the Rhine by 2020 (ICPR, 2004);
- the improvement of flood mitigation by implementing the Action Plan on Floods;
- the further improvement of water quality and groundwater protection;

- The continuous surveillance of the state of the Rhine and further improvement of water' (www.iksr.org/index.php?id=30&L=3).

Assessment

Known benefits[15]

- Improved water quality: 96% of the population are connected to a wastewater treatment plant; many big industrial plants have their own wastewater treatment plant. However, a few substances are still detected in too high concentrations in water or suspended matter (ICPR, *Rhine 2020: Programme on Sustainable Development, Balance 2000-2005:* www.iksr.org/index.php?id=160&L=3).
- Increased number of animal and plant species: 63 fish species now live in the Rhine. In particular, and since 2006, salmon (previously extinct), sea trout and eel as well as other migratory fish migrate from the North Sea as far upstream as Strasbourg.
- Flood prevention: additional flood retention areas have been created. Since the floods of 1995, almost all flood prevention measures planned to be achieved by 2005 have been implemented.
- Ecological improvements: floodplains have been reactivated; oxbow lakes have been reconnected with the Rhine and tributaries; and in many smaller sections the river bank structures have been ecologically improved.

Challenges (and when they exist, mechanisms to overcome them)

- Inadequate legal framework hampered initial efforts at co-ordination as it did not provide for rules or procedures for decision making. Stronger legal basis and greater institutional capacity introduced in 1999 has been instrumental in improving co-ordination (Huisman et al., 2000).
- Incongruence in preferences between upstream and downstream countries (e.g. Rhine chlorides dispute among Netherlands and the upstream Rhine riparian states that resulted in the Rhine Chlorides Convention of 1990) – this was finally addressed through agreements on the distribution of costs for prevention and remediation between the riparian states, but required arbitration, emphasizing the importance of an effective dispute resolution mechanism, agreed in advance (Dieperink, 2011).

- Establishment of a common system of data collection, measurement and analysis, largely through the ICHR, in order to establish a common basis for an objective assessment of water quality – this has taken several decades but it has facilitated joint monitoring and assessment exercises, created a common understanding of the issues and been important in forming the basis for the formulation of joint measures (Frijters and Leentvaar, 2003).
- Integration of emerging ecological challenges into the overall institutional and management framework – this is part of the Rhine 2020 programme.

Costs

The annual budgets of ICPR are not available. However, it is estimated that the Action Plan on Floods will cost EUR 12 billion by 2020 (ICPR, "Action Plan on Floods", www.iksr.org/index.php?id=123&L=3).

The Rules of Procedure and Financial Regulations stipulate that the costs of the ICPR are to be distributed as follows: the European Community (2.5%), the Swiss Confederation (12%), the remaining 85.5% is distributed in the following shares: Germany (32.5%), France (32.5%), Luxembourg (2.5%) and the Netherlands (32.5%).

Next steps

Future core areas of concern for the ICPR are the following (ICPR, 2010):

- Restore the biological continuity and increase habitat variety, i.e. the structures of river bottoms, river banks and alluvial areas;
- Further improve water quality, equally taking into account residues of pharmaceuticals, hormone, active ingredients etc;
- Further improve flood protection and flood prevention;
- Draft adjustment strategies to the effects of climate change.

Notes

1. Thus it does not focus on management of the marine environment; nor does it focus on the transboundary management of aquifers, which has recently come onto policy agendas, as highlighted by UNESCO and UNECE: World-wide Hydrological Mapping and Assessment Programme (WHYMAP), http://typo38.unesco.org/en/about-ihp/associatedprogrammes/whymap.html; Almássy and Buzás (1999); Puri, et al. (2001).

2. A recent study on floods in a transboundary context concluded that although only 10% of all river floods are transboundary, these floods represent a considerable amount of the total number of casualties, displaced/affected individuals and financial damages worldwide: Bakker (2006).

3. Of these conflicts, 32 took place in the Middle East, and practically all concerned the Jordan River and its tributaries (Belt, 2010).

4. An analysis by Wolf and Hamer (2000) of treaties dealing with non-navigational issues of water management, flood control, hydropower projects or allocations for consumptive or non-consumptive uses in international basin found that 124 of the 145 treaties (86%) are bilateral and 21 (14%) are multilateral (2 of which are unsigned agreements or drafts); of the latter, agreements between developing nations account for 13.

5. Lebanon and Syria have signed in 1994 the Accord Concerning the Distribution of the Orontes. The agreement does not involve Turkey due to a breakdown in negotiations: AFED Report, 2010.

6. States who have not ratified have expressed a number of concerns. Salman (2007) suggests that disagreement is based on a number of misconceptions about the Convention, notably that neither upper nor lower riparian states are content with the compromise between the principles of "reasonable and equitable use" and "no harm", each thinking the Convention favours the other, nor are upper riparians content with the requirement to notify the other of planned uses. Some of those with

existing agreements feel these are not adequately recognised under the Convention. Further, some argue that the dispute resolution mechanisms are too weak as they are non-binding; others argue that the fact finding commission is an inappropriate intrusion into their sovereignty. The position of countries can be inconsistent, however. Although the 14 members of the South African Development Corporation revised their own Protocol on Shared Watercourse Systems to bring it in line with the Convention, only 2 of those countries have ratified the convention itself.

7. See for instance the report to the French Parliament: "Rapport d'information déposé par la Commission des affaires étrangères sur la géopolitique de l'eau", October 2010: www.assemblee-nationale.fr/13/rap-info/i4070.asp.

8. Co-operation on shared waters between the US and Canada dates back to the 1909 Boundary Waters Treaty. Water disputes have traditionally been resolved through the International Joint Commission (www.ijc.org), and interesting dispute resolution mechanism where the US and Canada are equal (both countries appoint 3 commissioners). However, the evidence shows that water disputes are growing and increasingly settled through domestic courts. See Parrish (2005) and Hall (2006).

9. E.g. the Sandoz fire on the Rhine was the trigger for the Rhine Action Plan in 1987; flooding in 1995 enhanced co-operation with respect to both the Rhine and the Meuse: Huisman et al. (2000); however others argue that agreements forged before crisis situations emerge can be stronger: A. Wolf and J. Newton, "Case Study Transboundary Dispute Resolution: the Mekong Committee" www.transboundarywaters.orst.edu/research/case_studies/Documents/mekong.pdf.

10. In a recent study of the Mekong River Commission, Schmeier (2011) found that there were relatively weak linkages between the MRC and national governance structures, impeding its effectiveness.

11. Downstream countries can be compensated for the creation and operation of additional storage capacity or for pollution by upstream countries (Dinar, 2006); a mountainous country with hydropower but little arable land might trade the electricity and water for irrigation from its mountain streams for agricultural products from the irrigated country downstream.

12. A method of accounting is under development (Payment for Ecosystem Services, PES), but remains contested, though is seen by the UN as having the potential to offer considerable benefits to IRWM in the longer term (UN-Water, 2008; UNDP, 2004).

13. For example, Germany and Switzerland have recently given EUR 8 m and 2.4 m respectively to the Mekong River Commission to support projects on development and environmental protection: www.mrcmekong.org.

14. i.e.: maintaining and improving the quality of the Rhine's waters; preventing, reducing or eliminating pollution and ensuring and improving the safety of installations and preventing incidents and accidents; protecting populations of organisms and species diversity and reducing contamination; maintaining, improving and restoring the natural function of the waters and of alluvial areas as natural floodplains; conserving, improving and restoring habitats possible wild fauna and flora including migratory patterns for fish; ensuring environmentally sound and rational management of water resources; taking ecological requirements into account when implementing technical measures to develop the waterway, e.g. for flood protection, shipping or the use of hydroelectric power.

15. ICPR website, International Cooperation, "Success", www.iksr.org/index.php?id=151&L=3.

Annex 2.A1

International legal and soft law instruments affecting transboundary water governance

Legal instrument	Date	Description	Legal status
International			
Helsinki Rules on the Uses of International Watercourses (International Law Association) http://webworld.unesco.org/water/wwap/pccp/cd/pdf/educational_tools/course_modules/reference_documents/internationalregionconventions/helsinkirules.pdf	1966	Spells out rights and obligations for transboundary rivers, lakes and underground aquifers. Incorporates the principle of equitable utilisation, i.e. that each state within an international drainage basin has the right to a reasonable and equitable share of the beneficial use of the basin waters.	Non-binding statement on customary law issued by the International Law Association. Widely respected. Supplanted by Berlin Rules in 2005.
Convention on wetlands of international importance (Ramsar Convention) www.ramsar.org/	1971	Provisions for the ecological protection of wetlands	Binding on states who have ratified
United Nations Convention to Combat Desertification www.unccd.int/en/Pages/default.aspx	1994	Objective is to combat desertification and mitigate the effects of drought, particularly in Africa, through effective action at all levels, supported by international co-operation and partnership arrangements; includes provisions for the development of joint programmes for the sustainable management of transboundary natural resources, including water, through bilateral and multilateral mechanisms.	Binding from 1996 on states who have ratified
United Nations Convention on the Law of Non-navigable Uses of International Watercourses http://untreaty.un.org/ilc/texts/instruments/english/conventions/8_3_1997.pdf	1997	The only global international agreement covering the management and use of transboundary waters; provides a framework that states can build upon and implement bilaterally and multilaterally. Requires riparian states to co-operate in order to optimally use and adequately protect international watercourses, including through prior notification of planned activities, exchange of information and harmonisation of pollution policies. Dispute settlement either through ICJ or separate Fact Finding commission.	Requires 35 UN members to ratify; not yet binding but is a statement of customary international law (which is binding). Has formed the basis for several subsequent multilateral and bilateral agreements, e.g. on the Mekong and Danube rivers.

Legal instrument	Date	Description	Legal status
Berlin Rules on Water Resources Law (International Law Association www.ila.org).	2004	Revision to Helsinki Rules incorporating legal developments in international environmental and humanitarian law since 1966 as they apply to transboundary water management.	Non-binding statement of international customary law issued by the International Law Association
Regional			
UNECE Convention on the Protection and Use of Transboundary Watercourses and International Lakes (Helsinki Convention) www.unece.org/fileadmin/DAM/env/water/pdf/watercon.pdf	1992	Requires member states to pass legislation regulating point source wastewater disposal and to adopt water quality objectives and criteria consistent with the convention. Requires states to establish and implement co-ordinated programmes to monitor and assess transboundary water conditions, administered by Working Party on Water Problems (WPWP).	Ratified by 36 countries and EU. Binding from 1996. Provided the basis for bilateral and multilateral conventions e.g. 1994 Convention on the Co-operation for the Protection and Sustainable Use of the Danube River. 2003 amendment made it open for accession by all UN members.
United Nations Convention on Biological Diversity www.cbd.int/convention/text	1992	Aims to ensure the conservation and sustainable use of biological resources; seeks to mitigate or avoid adverse impacts on biological diversity and to prevent any threat to biological diversity. Emphasises the need to conduct environmental impact assessments so as to minimise damage to ecosystems, particularly river ecosystems.	Binding on all who have ratified
United Nations Economic Commission for Europe (UNECE) Convention on Environmental Impact Assessment in a Transboundary Context	1991	Provides that parties must assess the environmental impact activities with transboundary implications at an early stage of planning and lays down the general obligation of States to notify and consult each other on all major projects under consideration that are likely to have a significant adverse environmental impact across boundaries. Includes obligations regarding participation.	Entered into force 1997; amended by Kiev Protocol in 2004 which requires parties to undertake strategic impact assessments at an earlier stage than the EIAs required under the Convention.
United Nations Economic Commission for Europe (UNECE) Protocol on Water and Health www.unece.org/env/water/text/text_protocol.html	1999	Builds on the 1992 Convention. Covers the prevention, control and reduction of water-related diseases in Europe. Requires member states to establish national/local targets for drinking water quality, discharge quality, water supply performance and wastewater treatment to reduce water-related disease outbreaks. Emphasises need for international co-operation and action, exchange of information and knowledge about water management problems and risks	Binding on all who have ratified

Legal instrument	Date	Description	Legal status
UNECE Convention on Access to Information, Public Participation in Decision-Making and Access to Justice in Environmental Matters (Aarhus Convention) www.unece.org/fileadmin/DAM/env/pp/documents/cep43e.pdf	1999, adopted in 2001	Creates rights of public access to environmental information; rights of public participation in the making of environmental decisions; and access to justice in environmental matters	Binding on all who have ratified; incorporated into EU law and adopted by the EU institutions. Relevant to the environmental aspects of transboundary water management.
UNECE Guidelines on Sustainable Flood Prevention (UNECE, 2000) www.unece.org/fileadmin/DAM/publications/oes/Transboundary_Flood_Risk_Management_Final.pdf	2000	Guidelines on the prevention, control and reduction of transboundary impacts from floods and unilaterally decided flood protection measures such as dams	Supplements 1997 Convention; non-binding
UNECE Guidelines on Monitoring and Assessment of Transboundary Rivers, Lakes and Groundwaters www.unece.org/fileadmin/DAM/env/water/publications/documents/StrategiesM&A.pdf	2006	Intended to harmonise monitoring and assessment systems and require signatories to develop co-ordinated implementation of water policies based on sound institutional arrangements that facilitate co-operation.	Supplements 1997 Convention; non-binding
UNECE Recommendations on payments for ecosystem services (PES) www.unece.org/fileadmin/DAM/trans/osce/osceunece/misc_Zaragoza_Recommendations.pdf	2007	Provide guidance on the establishment and use of PES to implement integrated water resources management through the promotion of the protection, restoration and sustainable use of water-related ecosystems at all levels, from local to transboundary, and to promote the development of a market for ecosystem services	Supplements 1997 Convention; non-binding
UNECE Guidance on Water and Adaptation to Climate Change www.unece.org/fileadmin/DAM/env/water/publications/documents/Guidance_water_climate.pdf	2009	Advice to decision makers on how to assess impacts of climate change on water quantity and quality, how to perform risk assessment, including health risk assessment, how to gauge vulnerability, and how to design and implement appropriate adaptation strategies, focusing on transboundary issues.	Guidance
European Union Water Framework Directive 2000 http://ec.europa.eu/environment/water/water-framework/index_en.html	2000	Covers all EU member states' surface and groundwater resources, both domestic and transboundary. Introduces notion of integrated management of river basins. Combines source controls with water quality requirements. Introduces requirement for water prices to reflect costs. Requires riparian states to co-operate in the management of transboundary river basins.	Binding on EU member states.

Legal instrument	Date	Description	Legal status
European Union Groundwater Directive http://ec.europa.eu/environment/water/water-framework/groundwater/policy/current_framework/new_directive_en.htm	2006	Implements the provisions of the EU Water Framework Directive on the prevention and control of groundwater pollution. Sets criteria to assess groundwater chemical status and to identify pollution trends. Regulates input of pollutants into aquifers; and fights deterioration of all groundwater bodies	Binding
African Convention on the Conservation of Nature and Natural Resources http://au.int/en/sites/default/files/african_convention_conservation_nature_and_natural_resources.pdf		Commits parties to obligations to consult on the use of transboundary waters and if necessary establish an inter-State commissions to study and resolve problems arising from their joint use, development and conservation.	States report to Organisation on African Unity

Annex 2.A2

Examples of international agreements for the transboundary management of individual water courses

River basin	Agreement(s)	Date(s)	Institutional structure	Members
Rhine	Agreement concerning the International Commission for the Protection of the Rhine against Pollution.	1963, revised in 1976 and 1999	International Commission for the Protection of the Rhine	Founding states: Switzerland, the Netherlands, France, Germany, Luxemburg. European Economic Community (EEC) in 1976
Danube	Convention concerning the regime of navigation on the Danube Convention Concerning Fishing in the Waters of the Danube Convention on the Co-operation for the Protection and Sustainable Use of the Danube River	1948 1958 1994	Danube Commission Mixed Commission on fishing International Commission for the Protection of the Danube River	Union of Soviet Socialist Republics, Bulgaria, Czechoslovakia, Hungary, Romania, Ukraine Soviet Socialist Republics and Yugoslavia Romania, Bulgaria, Yugoslavia and the Union of Soviet Socialist Republics Germany, Austria, Czech Republic, Croatia, Slovakia, Hungary, Slovenia, Bosnia-Herzegovina, Serbia, Romania, Bulgaria, Moldova, Ukraine, Montenegro, EU

River basin	Agreement(s)	Date(s)	Institutional structure	Members
Nile	Exchange of Notes between Her Majesty's Government in the United Kingdom and the Egyptian Government on the Use of Waters of the Nile for Irrigation Agreement between the Republic of the Sudan and the United Arab Republic for the full utilisation of the Nile waters. Nile Basin Initiative Co-operative Framework Agreement	1929 1959 1999 2010	Establishment of property rights among Egypt and Sudan, which was a UK's protectorate. No new institutional structures. Joint technical commission. Council of ministers of water affairs of the Nile Basin (Nile-COM), a technical Advisory Committee (Nile-TAC), and a secretariat (Nile-SEC) Establishes the Nile River Basin. Commission comprised of: (a) Conference of Heads of State and Government (b) Council of Ministers; (c) Technical Advisory Committee; (d) Sectoral Advisory Committees; (e) Secretariat.	Egypt and United Kingdom Sudan and UAE Egypt, Sudan, Burundi, Democratic Republic of Congo, Eritrea (observer), Ethiopia, Kenya, Rwanda, Tanzania and Uganda Ethiopia, Kenya, Uganda, Rwanda and Tanzania were original signatories with Burundi signing in February 2011; Egypt, Sudan and Republic of Congo have not yet signed
Niger Basin	Agreement concerning the Niger River Commission and the navigation and transport on the River Niger Convention creating the Niger Basin Authority (with protocol relating to the development fund of the Niger Basin)	1964 1980	Niger River Commission Niger Basin Authority	Cameroon, Ivory Coast, Dahomey, Guinea, Upper Volta, Mali, Niger and Chad Niger, Benin, Chad, Guinea, Ivory Coast, Mail, Nigeria, United Republic of Cameroon and Upper Volta
Lake Chad	Fort Lamy (N'Djamena) Convention	1964	Lake Chad Basin Commission	Cameroon, Niger, Nigeria, CAR, Chad
Gambian River Basin	Convention Relating to the Creation of the the Gambia River Basin	1967	Gambian River Basin Development Organisation	The Gambia, Guinea, Guinea Bissau, Senegal
Lake Victoria	Convention for the Establishment of the Lake Victoria Fisheries Organisation Protocol for the Sustainable Development of Lake Victoria Basin	1994 2003	Lake Victoria Fisheries Organisation Establishes Lake Victoria Basin Commission	Kenya, Uganda and Tanzania Kenya, Uganda and Tanzania

River basin	Agreement(s)	Date(s)	Institutional structure	Members
Volta Basin	Convention on the status of the Volta River and the Establishment of Volta Basin Authority	2007	Volta Basin Authority	Burkina Faso, Benin, Côte d'Ivoire, Ghana, Mali, Togo
Southern African Development Community	Protocol on Shared Watercourse Systems in the Southern African Development Community Region	1995, revised and extended in 2000	Establishes the following SADC Water Sector Organs: *i)* the Committee of Water Ministers; *ii)* the Committee of Water Senior Officials; *iii)* the Water Sector Co-ordinating Unit; and *iv)* the Water Resources Technical Committee and sub-Committees.	Angola, Botswana, Congo, Lesotho, Malawi, Mauritius, Mozambique, Namibia, Seychelles, South Africa, Swaziland, Tanzania, Zambia, Zimbabwe
Mekong	Statute for the Committee for Co-ordination of Investigations into the Lower Mekong Basin Joint Declaration of Principles for the Utilisation of the Water of the Lower Mekong Basin Agreement on the Interim Committee for Co-ordination of Investigations into the Lower Mekong Basin; Agreement on the Co-operation for the Sustainable Development of the Mekong River Basin	1957 1975 1978 1995	Establishment of the Mekong Committee supported by the Mekong Secretariat; status of a regional UN body (1957) Mekong Committee Interim Committee Creation of the Mekong River Committee	Cambodia, Laos, Thailand, and Vietnam Cambodia, Laos, Thailand and Vietnam Laos, Thailand and Vietnam Cambodia, Laos, Thailand and Vietnam
Halala River, Kerulen River, Bor Nor Lake and Bulgan River	Agreement between China and Mongolia on the protection and utilisation of transboundary waters	1994	Create the Joint Committee on Transboundary Waters Establishes joint research and experiment centers or groups.	China and Mongolia
Mahakali River	Treaty Between Nepal and India Concerning The Integrated Development of the Mahakali Barrage Including Sarada Barrage, Tanakpur Barrage and Pancheshwar Project	1996	Mahakali River Commission.	India and Nepal
Ganges River	Treaty between India and Bangladesh on Sharing of the Ganga/Ganges Waters at Farakka	1996	Establishes a Joint Committee for implementing the arrangements of the Treaty.	India and Bangladesh

River basin	Agreement(s)	Date(s)	Institutional structure	Members
Indus	Indus Waters Treaty between India and Pakistan Indus Basin Development Fund Agreement The Indus Basin Development Fund Agreement	1960 1960 1964	Fixes and delimits rights and obligations concerning the use of the waters (e.g. exchange of data, settlement of disputes, emergency provisions). No new institutional structure. Establishes the Indus Basin Development Fund Establishes supplemental contributions to the Fund.	India and Pakistan Australia, Canada, Germany, New Zealand, Pakistan, United Kingdom and United States and the IBRD. Australia, Canada, Germany, New Zealand, Pakistan, United Kingdom and United States and the IBRD.
Cuareim River Basin	Agreement between Uruguay and Brazil for the use of natural resources and the development of the basin of the Cuareim river Complementary Adjustment to the Agreement between Brazil and Uruguay for the Utilisation of the Natural Resources and the Development and the Cuareim River Basin	1991 1997	Creates the Mixed Uruguayan-Brazilian Commission for the Development of Cuareim River Basin (CRC) Water allocation specifications. No new institutional structure.	Brazil and Uruguay

Sources: International Water Law Project (http://internationalwaterlaw.org/documents); Lake Chad Basic Commission Perspectives (www.oieau.fr/ciedd/contributions/atriob/contribution/cblt.htm); Mekong River Commission (www.mrcmekong.org); Nile Basin Initiative (www.nilebasin.org/newsite); Organization for the Development of Senegal River (www.omvs.org/fr/omvs/presentation.php); Office International De L'eau (www.oieau.fr/spip.php?article1181); Institute For Water And Watersheds – Oregon State University (www.transboundarywaters.orst.edu/database/interfreshtreatdata.html); International Joint Commission Canada-US (www.ijc.org/en/background/ijc_cmi_nature.htm) and Pakistan.Com (www.pakissan.com/english/watercrisis/the.water.accord.shtml).

Annex 2.A3

United Nations engagement

UN body	Key role	Members
UN-Water	An inter-agency mechanism created in 2002 which pulls together the main UN-related agencies, departments and programmes involved in water-related issues. It is responsible for follow-up to the water-related decisions reached at the 2002 World Summit on Sustainable Development and the Millennium Development Goals. It supports Member States in their efforts to achieve water- and sanitation-related goals and targets.	27 members including UNESCO, UNDP, the regional UN bodies, the Global Environment Fund, the World Health Organisation and the World Meteorological Organisation.
UN-Water Task Force on Transboundary Waters	• The Taskforce: • provides a platform to promote coherence and co-ordination of activities by UN-Water members and partners in the area of transboundary waters. • acts as an intermediary between the UN initiatives and practices at regional or local levels, acting acts as a "clearing house" for good-practices: facilitating a steady exchange of information, experiences and lessons learned and by promoting joint efforts to help identify gaps in programmes, maximise their effectiveness and avoid unnecessary duplication of effort. • aims to ensure that issues relating to transboundary waters occupy a high position on the political agenda • encourages collaboration between stakeholders.	United Nations Development Programme (UNDP) Global Environment Facility (GEF) United Nations Environment Programme (UNEP) International Atomic Energy Agency (IAEA) United Nations Economic and Social Commission for Asia and the Pacific (UNESCAP) UN-Water Decade Programme on Capacity Development (UNW-DPC) United Nations Office to support the International Decade for Action "Water for Life" (UNO-IDfA) World Meteorological Organization (WMO) World Health Organization (WHO) Convention on Biological Diversity (CBD) United Nations Convention to Combat Desertification (UNCCD) RAMSAR Convention United Nations University (UNU) Stockholm International Water Institute (SIWI)

UN body	Key role	Members
UN regional bodies	Engaged in supporting co-operation in the management of transboundary water courses in their regions. Notably, the United Nations Economic Commission for Asia and the Pacific (ESCAP) founded the original Mekong River Committee in 1957, and continues to support the Commission established in 1995.	United Nations Economic and Social Commission for Western Asia (ESCWA) United Nations Economic Commission for Africa (ECA) United Nations Economic Commission for Asia and the Pacific (ESCAP) United Nations Economic Commission for Europe (UNECE) United Nations Economic Commission for Latin America and the Caribbean (ECLAC).
Global Environment Facility	International financing mechanism established in 1991 to address global environmental issues. It provides assistance to developing and transition economies to improve cross-sectoral management of transboundary basins and aquifers by helping them to establish priorities, adopt policy legal and institutional reforms in sectors facing degradation or conflicts, and test the feasibility of various investments to address conflicts and reverse degradation. It assists in the development and strengthening of multi-country river basin institutions, fostering policy and legislative reforms, and promoting broad stakeholder involvement in addressing key threats and priorities. It has allocated USD 9.2 billion, supplemented by more than USD 40 billion in co-financing, for more than 2 700 projects in over 165 countries.	Partners with several UN agencies and programmes including UNDP; UNEP; the World Bank; FAO; UNIDO; the African Development Bank; the Asian Development Bank; the European Bank for Reconstruction and Development; the Inter-American Development Bank; and the International Fund for Agricultural Development. In 2011 established accreditation scheme for other agencies.

Sources: UN-Water (2008), "Transboundary waters: sharing benefits, sharing responsibilities", UN-Water Thematic Paper; UNDP (2004), "Water Governance for Poverty Reduction: Key Issues and the UNDP Response to Millenium Development Goals", United Nations Development Program, New York.

Annex 2.A4

Non-governmental organisations and initiatives on transboundary water

A number of non-governmental organisations are engaged in promoting co-ordination in water governance, including:

- **The Global Water Partnership (GWP)**: www.gwp.org

GWP aims to promote integrated water resource management through capacity building programmes, access to global and regional networks, resource centres and training materials, and exchange of information and current practices between countries. For example, GWP has been leading since 2000 the *Dialogues on Effective Water Governance* with the International Council for Local Environment Initiatives (ICLEI). GWP has also been running with UNDP and UNESCO-IHE (the Institute for Water Education) the Capacity Building Network for Integrated Water Resources Management (Cap-Net), a global programme of capacity building for the sustainable development of water resources.

- **The World Water Council (WWC)**: www.worldwatercouncil.org

The World Water Council was established in 1996 and acts as an international think thank on water management and to stimulate policy dialogue. The Council has over 300 active members from 60 countries. It aims to promote awareness, build political commitment and trigger action on critical water issues, notably through the World Water Forums which are held every three years. The WWC has a Working Group on Transboundary Governance which is tasked with preparing a series of perspective papers documenting current practices, progress and prospects in transboundary water management and wider regional co-operation, including how such practises contribute to economic growth, safeguard the environment and strengthen regional security. Building on these papers the WWC intends to prepare an analysis describing ways of overcoming political-economic obstacles to greater regional benefit sharing (WWC, A New Water Politics).

- **The International Network of Basin Organisations (INBO)**: www.inbo-news.org.

INBO was established in 1994 as a network of organisations whose common goal is to implement integrated basin water resource management. To date there are some 83 member organisations from some 34 countries.

- **Other International non-governmental organisations** (INGOs):

These include Green Cross (www.gci.ch), an international NGO headed by former world statesmen has as its objective conflict mitigation – including support to the resolution of international water disputes; the International Water Association (www.iwahq.org) and the International Union for Conservation of Nature (www.iucn.org).

Annex 2.A5

Development of co-ordination with respect to the Rhine: Main landmarks

1950: Upon an initiative of the Netherlands, the Rhine bordering countries Switzerland, France, Luxemburg, Germany and the Netherlands create a common forum for discussing and solving problems related to the pollution and restoration of the Rhine." (ICPR Website)

1963: With the signing of the agreement on the International Commission on the Protection of the Rhine against Pollution (Berne Convention) the Rhine bordering states create a basis under international law for their co-operation.

1976: The European Economic Community signs a supplementary agreement to the Berne Convention and becomes party to this convention.

1976: Signing of the Convention on the protection of the Rhine against Chloride Pollution (Chlorides Convention), a supplementary agreement to which was signed in 1991.

1976: Signing of the Convention on the Protection of the Rhine against Chemical Pollution.

November 1986: Sandoz accident and two meetings of ministers determining the guidelines for the Rhine Action Programme.

1987: Ministers decide on the implementation of the Rhine Action Programme.

1995: Ministers require the drafting of an "Action Plan on Floods".

1998: Conference of Rhine ministers (22.01.98) passing the Action Plan on Floods and adopting the text of the new Rhine Convention.

2000: The EC Water Framework Directive (2000/60/EC) enters into force.

2001: Conference of Rhine ministers (29.01.01) – "Rhine 2020" adoption.

2003: The Convention on the Protection of the Rhine enters into force.

2006: The EC-Groundwater directive 2006/118/EC enters into force.

2007: Entry into force of the EC Floods directive (2007/60/EC).

Bibliography

Almássy, E. and Zs. Buzás (1999), *Inventory of transboundary groundwaters*, Vol. 1, UN/ECE Task Force on Monitoring and Assessment, Lelystad, The Netherlands.

Bakker, M. (2006), *Transboundary river floods: Vulnerability of continents, international river basins and countries*, Oregon State University, Oregon, US.

Belt, D. (2010), "Parting the Waters", in *National Geographic*, April, http://ngm.nationalgeographic.com/2010/04/parting-the-waters/belt-text.

Bernauer, T. and P. Moser (1996), "Reducing pollution of the Rhine river: The influence of international cooperation", *Journal of Environment and Development*, Sage Journals, Vol. 5, pp. 389-415.

Dieperink, C. (2011), "International water negotiations under asymmetry, Lessons from the Rhine chlorides dispute settlement (1931–2004)", *International Environmental Agreements: Politics, Law and Economics*, Springer, Vol. 11, No. 2, pp. 139-157, May.

Dinar, S. (2006), "Assessing side-payments and cost-sharing patterns in international water agreements: The geographic and economic connection", *Political Geography*, Vol. 25, Issue 4, pp. 412-437, DOI:10.1016/j.polgeo.2006.03.007, May.

ECOWAS – SWAC/OECD (2006), "Transboundary river basins" in *Atlas on Regional Integration in West Africa, Land Series*, www.oecd.org/dataoecd/41/45/38409569.pdf.

Amer, E. D. et al. (2005), "Sustainable development and international cooperation in the Eastern Nile Basin", *Aquatic Sciences*, Vol. 67, pp. 3-14.

Frijters, I. and J. Leentvaar (2003), "Rhine case study", International Hydrological Programme, UNESCO, Paris.

Hall, N. D. (2006), "Bilateral Breakdown, U.S. – Canada Pollution Disputes", 21 Natural Resources & Environment 18, Summer, http://works.bepress.com/noahhall/3.

Huisman, P, J. Jong and K. Wieriks (2000), "Transboundary cooperation in shared river basins: experiences from the Rhine, Meuse and North Sea", *Water Policy* Vol. 2, pp. 83-97.

ICPR (2010), "Living Rhine, 60 years ICPR", International Commission for the protection of the Rhine, www.iksr.org/fileadmin/user_upload/Dokumente_en/Brochures/IKSR-Image_60Jahre_uk_web__2_.pdf.

ICPR (2004), "Rhine Salmon 2020", International Commission for the protection of the Rhine, www.iksr.org/index.php?id=124&L=3.

ICPR (2004), "Rules of procedure and financial regulations", www.iksr.org/fileadmin/user_upload/Dokumente_en/Gesch_fts-Finanzordnung-e.pdf.

Jacobs, J. (2002), "The Mekong River Commission: transboundary water resources planning and regional security", *The Geographical Journal,* Vol. 168, No. 4.

Jacobs J. (1995), "Mekong Committee History and Lessons for River Basin Development", *The Geographical Journal*, Vol. 161, No. 2.

Joyce J. and J. Granit (2010), "Challenges and opportunities for financing transboundary water resources management and development in Africa", Concept Note, EUWI.

Kliot, N., D. Shmueli and U. Shamir (2001), "Institutions for management of transboundary water resources: their nature, characteristics and shortcomings, *Water* Policy, Vol. 3, Issue 3, pp. 229–55.

Komakech C., F. Jaspers, P. van der Zaag (2007), "Strengthening transboundary river basin institutions: a comparative learning from basin institutions in Africa, Europe and Southeast Asia", UNESCO.

Mostert, E. (2003), "Conflict and co-operation in the management of international freshwater resources: A global review", RBA Centre, Delft University of Technology, Delft, The Netherlands.

Newton, J. (n.d.), "Case Study of transboundary dispute resolution: organization for the development of the Senegal river (OMVS)", Oregon State University, www.transboundarywaters.orst.edu/research/case_studies/Documents/senegal.pdf.

ODI (2001), *Transboundary Water Management as an International Public Good, Development Financing 2000*, Swedish Ministry for Foreign Affairs, www.odi.org.uk/resources/docs/3870.pdf.

OECD (2011), *Water Governance in OECD Countries: A Multi-level Approach*, OECD Publishing, Paris, www.oecd.org/gov/water, doi: 10.1787/9789264119284-en.

OECD (2009), *Managing Water for All: An OECD Perspective on Pricing and Financing*, OECD Studies on Water, OECD Publishing, Paris, doi:10.1787/9789264059498-en, www.oecd.org/water.

Parrish, A. L. (2005), "Trail Smelter Deja Vu: Extraterritoriality, International Environmental Law, and the Search for Solutions to Canada-U.S. Transboundary Water Pollution Disputes", *Boston University Law Review*, Vol. 85, No. 2, http://ssrn.com/abstract=1019924.

Puri, S. et al., (2001), "Internationally shared (transboundary) aquifer resources management: their significance and sustainable management", A framework document, Non Serial Document SC-2001/WS/40, UNESCO, Paris.

UNESCO (2003), "Water for People, Water for Life", United Nations World Water Development Report 1, Paris.

Salman (2007), "The UN Water Courses Convention Ten Years On: Why Has Entry into Force Proven Difficult?", *Water International*, Vol. 32, No. 1, pp. 1-15, World Bank, Washington DC.

Savenije, H. G. and P. Van Der Zaag (2000), "Conceptual Framework for the Management of Shared River Basins, With Special Reference to the SADC and EU", *Water Policy*, Vol. 2, pp.9–45.

Schmeier, S. (2011), "Resilience to climate change-induced challenges in the Mekong River Basin – the role of the MRC", *Water Papers*, World Bank, Washington DC.

Schwabach A. (1989), "The sandoz spill: the failure of international law to protect the Rhine from pollution", *Ecology Law Quaterly*, Vol. 16, p. 443.

Stephan, R. M. (2010), "Trans-Boundary Water Resources", Chapter 10, in *Arab Environment Water: Sustainable Management of a Scarce Resource,* 2010 Report of the Arab Forum for Environment and Development (AFED), www.afedonline.org/Report2010/pdf/En/Chapter10.pdf.

Timmerman, J. and F. Bernardini (2007), "Adapting to climate change in transboundary water management", World Water Council.

UNDP (2004), "Water Governance for Poverty Reduction: Key Issues and the UNDP Response to Millenium Development Goals", United Nations Development Program, New York.

UN-Water (2008), "Transboundary waters: sharing benefits, sharing responsibilities", UN-Water Thematic Paper.

Wolf, A. T. (1997), "International water conflict resolution: Lessons from comparative analysis", *International Journal of Water Resources Development,* Vol. 13, Issue 3, pp. 333-366, DOI:10.1080/07900629749728.

Wolf, A. T. and J. H. Hamer (2000), "Trends in Transboundary Water Disputes and Dispute Resolution", in *Water for Peace in the Middle East and Southern Africa* prepared by Green Cross International for the 2nd World Water Forum in The Hague, March, Geneva: Green Cross International.

Wolf, A. T. et al. (1999), "International River Basins of the World", *International Journal of Water resources Development*, Vol. 15, No. 4, December.

Wolf A. T. and J. Newton (n.d.), "Case Study Transboundary Dispute Resolution: the Mekong Committee", www.transboundarywaters.orst.edu/research/case_studies/Documents/mekong.pdf.

ORGANISATION FOR ECONOMIC CO-OPERATION AND DEVELOPMENT

The OECD is a unique forum where governments work together to address the economic, social and environmental challenges of globalisation. The OECD is also at the forefront of efforts to understand and to help governments respond to new developments and concerns, such as corporate governance, the information economy and the challenges of an ageing population. The Organisation provides a setting where governments can compare policy experiences, seek answers to common problems, identify good practice and work to co-ordinate domestic and international policies.

The OECD member countries are: Australia, Austria, Belgium, Canada, Chile, the Czech Republic, Denmark, Estonia, Finland, France, Germany, Greece, Hungary, Iceland, Ireland, Israel, Italy, Japan, Korea, Luxembourg, Mexico, the Netherlands, New Zealand, Norway, Poland, Portugal, the Slovak Republic, Slovenia, Spain, Sweden, Switzerland, Turkey, the United Kingdom and the United States. The European Union takes part in the work of the OECD.

OECD Publishing disseminates widely the results of the Organisation's statistics gathering and research on economic, social and environmental issues, as well as the conventions, guidelines and standards agreed by its members.

OECD PUBLISHING, 2, rue André-Pascal, 75775 PARIS CEDEX 16
(42 2013 17 1 P) ISBN 978-92-64-20051-7 – No. 60687 2013-01

IWA Publishing's authorised EU representative for General Product Safety Regulations is Diane D'Arras, 15 rue Duret, 75116 Paris, France, e-mail: safety@iwap.co.uk.

Printed and bound by CPI Group (UK) Ltd, Croydon, CR0 4YY

16/06/2026

02140366-0001